Build Your Own Web Framework in Elixir

Develop lightning-fast web applications using Phoenix and metaprogramming

Aditya Iyengar

BIRMINGHAM—MUMBAI

Build Your Own Web Framework in Elixir

Group Product Manager: Rohit Rajkumar
Publishing Product Manager: Aaron Tanna
Senior Editor: Mark D'Souza
Senior Content Development Editor: Feza Shaikh
Technical Editor: Saurabh Kadave
Copy Editor: Safis Editing
Project Coordinator: Manthan Patel
Proofreader: Safis Editing
Indexer: Pratik Shirodkar
Production Designer: Jyoti Chauhan
Marketing Coordinator: Anamika Singh, Namita Velgekar, and Nivedita Pandey

First published: June 2023

Production reference: 1170523

Published by Packt Publishing Ltd.
Livery Place
35 Livery Street
Birmingham
B3 2PB, UK.

ISBN 978-1-80181-254-2

www.packtpub.com

To my wife, Susan Walker, and my parents, Nirmal and Madhu Iyengar, without whose encouragement and support I wouldn't have even thought of writing a book.

– Aditya Iyengar

Foreword

I first met Adi when he applied to work at Annkissam in 2015. I was the **Chief Technology Officer** (**CTO**) at the time, but I wasn't involved in his formal interview process. It was fortunate that I saw him between interviews and we chatted about **error correction code** (**ECC**) memory and Hamming codes. I came away excited and had strong opinions that I needed to share with the hiring team. Thankfully, they agreed, and soon after, I had the opportunity to manage and mentor a brilliant developer.

At that time, our organization was running into concurrency issues with Ruby. I had looked into several technologies, but the Phoenix framework was the most impressive. It had just reached its 1.0 release, and with it came the benefits of Elixir and **Open Telecom Platform** (**OTPs**). We trained Adi on Rails for his first few weeks, but within a month, he helped me build a Phoenix clone of a Rails app, and its speed was impressive. We launched our first Phoenix application into production that same year.

Coming from **Ruby on Rails** (**RoR**), we lost a lot of convenient gems. Elixir was a newer ecosystem, but we were tech evangelists. The solution was to build everything we needed and open source as much as possible. No pagination module? Adi built Rummage. Our deployment strategy needed updating. Adi built Akd. DevOps? I used his tool to configure my last several laptops. My guidance during that time was to write tests, predict where other developers would have difficulty, and always read the code you're using.

It's that foundation that I see in this book and why I was so humbled that Adi asked me to write this foreword. While Phoenix is ubiquitous today, most guides focus on how to use it. In this book, you'll learn how it is built. You'll cover not just the fundamentals of web development but also how they're implemented. You'll learn how the framework has such elegant syntax and how you can use metaprogramming to do the same in your code. You'll also be shown plenty of examples to put it into context and ensure you can incorporate them into your own work.

The attention to detail you'll find in this book extends to its very publishing. Adi wanted to ensure the code samples were accurate, so he wrote unit-tested code, annotated that code, and then wrote a library to extract that code and combine it with the text to produce the final copy. He's always talking about 100% code coverage, so it's fitting he's extended that to printed work. It's that level of care which makes him a great mentor, and you'll find it throughout this book.

What I'm most excited about is that this book reminds us to continue asking how. Going deeper than the README.md file and hexdocs will make you a better developer. With Adi's guidance, this book does that for one of the most significant libraries in Elixir.

Eric Sullivan

Senior Software Engineer at PepsiCo, Ex-CTO, Ex-Founder

Contributors

About the author

Adi Iyengar is a staff software engineer who has worked with Elixir since 2015. Over those years, he has worked across a wide array of applications and authored/contributed to several open source projects, including Elixir itself. Adi has also advised and continues to advise several start-ups on engineering and product, and helps them adopt Elixir and Phoenix to build functional and scalable **minimal viable products** (**MVPs**) from the ground up. He is passionate about mentoring and sharing his knowledge with others, which is why he actively mentors other engineers. He loves Elixir, functional programming, and **test-driven development** (**TDD**). Adi is also the co-host of the Elixir Mix podcast. When not coding, Adi can be seen playing billiards, playing the guitar, or breakdancing. Adi also spends a good amount of his time keeping up with new developments in particle physics.

Adi is also the co-host of the Elixir Mix podcast which is a weekly Elixir podcast where the panel discusses Functional Programming, the Elixir ecosystem and building real-world applications using Elixir based tools and frameworks.

I have been extremely lucky in getting a lot of opportunities to learn and grow as an engineer, and I'd like to thank everyone who has contributed to my career, including my family, friends, mentors, and mentees. Thanks to every engineer who has contributed to this book in one way or another (in no particular order): Susan Walker, Eric Sullivan, Jeffrey Matthias, Josh Scott Adams, Sophie Debenedetto, Tuomo Hopia, Bruce Tate, Evan Kaplan, Andrea Leopardi, Sascha Wolf, Allen Wyma, and Charles Max Wood. Lastly, special thanks to a special company, Annkissam, because of whom I started walking the path to learning this amazing language.

About the reviewers

Henry Clinton is a versatile engineer with extensive experience in electronics and communication. He has honed his skills as a senior Elixir developer, delivering end-to-end execution of several pivotal projects at Iteron Technologies. Henry's passion for technology is infectious, and his insatiable curiosity drives him to explore and master a wide range of topics. He combines his technical expertise with a creative mindset to come up with innovative solutions to complex problems. His attention to detail and commitment to excellence is unparalleled, making him a valuable asset and instrumental in bringing his work to fruition.

Lubien is a software developer who specializes in Elixir and JavaScript languages. Previously focused on frontend work with JavaScript frameworks, he's now looking forward to technologies that handle the frontend on the backend, such as LiveView and LiveWire. He is passionate about teaching the latest trends in the industry. He enjoys working with Devs Norte to build a developer community in his hometown of Belém do Pará. He also posts content on his YouTube channel in Portuguese, @lubiendev, to help newcomers to Phoenix LiveView and writes English content under Fly.io Phoenix Files.

Nikhil More is a senior software developer with over nine years of experience. He has working experience in the domains of banking and finance, defense, logistics, mobility, and **internet of things (IoT)**. Prior to working on Elixir since 2016, he also worked on C++, .Net, PHP, JSP, and so on. His current interests include the Semantic Web, IoT, project management, Agile, and artificial intelligence (deep learning and genetic algorithms). Nikhil has taught deep learning at Usha Mittal Institute of Technology, Mumbai. He is currently working on his own software consultancy called Infoquarry Technologies Pvt. Ltd. Some of his previous notable employers are Electronics and Radar Development Establishment, Bengaluru, which is a premier laboratory of the Ministry of Defence, India, Western Railway, under the Ministry of Railways, India, Central Bank of India, BlockFi Inc., Volansys Technologies Pvt. Ltd, and Ahmedabad. Other mentions of the projects he has worked on are for a car-sharing application owned by a large Japanese automobile manufacturer and a Wi-Fi mesh device for a large US entity. He is also an active member of various technical meetup communities in Ahmedabad, Bengaluru, Mumbai, and Pune. He also conducts training sessions on functional programming and Elixir development.

Parvatham Raghuchander is an experienced enterprise cloud solutions architect with expertise in messaging and **voice-over IP (VoIP)**. She has developed, designed, re-engineered, and integrated enterprise solutions for a wide range of organizations. Her programming experience is mostly in Erlang, Elixir, and Golang, with specializations in Phoenix, the Cowboy framework, AWS, Azure, Docker, Kubernetes, NoSQL, and SQL databases.

Table of Contents

Part 1: Web Server Fundamentals

1

Introducing the Cowboy Web Server   3

2

Building an HTTP Server in Elixir   27

Part 2: Router, Controller, and View

3

Defining Web Application Specifications Using Plug   57

4

Working with Controllers   73

5

Adding Controller Plugs and Action Fallback   91

6

Working with HTML and Embedded Elixir   109

7

Working with Views   127

Part 3: DSL Design

8

Metaprogramming – Code That Writes Code   151

9

Controller and View DSL   187

10

Building the Router DSL   221

Index   249

Preface

There's an old story about the person who wished his computer were as easy to use as his telephone. That wish has come true since I no longer know how to use my telephone.

– Bjarne Stroustrup

The preceding quote from Stroustrup shows just how fast technology is evolving. We're building more complex software day by day while also building it at an ever-increasing rate. However, as the complexity of software increases, so do abstractions around its fundamentals. As a result, it is easier to build software using those abstractions but harder to completely understand how a piece of software works the way it does. This is true for programming languages, deployment tools, and several other types of tools we use. Another one of those tools is a web framework.

A web framework is a tool or a set of tools that assists in building and releasing a web application. This is accomplished through the process of standardization and building abstractions around those standardized models. Web frameworks usually include tools for server management, data management, templating, and so on. From the 1990s to 2021, we have seen the evolution of web frameworks from Java Servlets and ASP.NET to Rails and Elixir's Phoenix. Over the course of this period, we have significantly decreased the average amount of effort required to build a web application. However, since Phoenix facilitates productive web development, I frequently come across developers being bewildered by its "magic." That "magic" is often one obstacle that stands in the way of an Elixir web developer to cross the seemingly never-ending bridge of mid-level to senior-level. It is also one of the key factors that determines how confident they feel as a developer.

This book aims to help developers overcome that obstacle by building a web framework from scratch using Elixir. The goal is to demystify the aforementioned Phoenix magic by breaking it down into components and designing/building them from the ground up while testing their expected behaviors. I expect developers to feel more confident in their web development skills and Elixir knowledge after reading this book. Maybe, some will even go on to make contributions to the Elixir open source community.

Who this book is for

This book is for web developers looking to understand how Elixir components are used in the Phoenix framework. Basic knowledge of Elixir will be useful to understand the concepts covered in the book more effectively.

What this book covers

Chapter 1, Introducing the Cowboy Web Server, covers how an HTTP server is designed and the details of Cowboy, the most used web server in the Elixir ecosystem. It also covers how to use Cowboy to build a simple web application that serves HTML. Finally, it details how to test the web application built using Cowboy.

Chapter 2, Building an HTTP Server in Elixir, uses the conclusions from *Chapter 1* and extends them to build a brand new HTTP server in Elixir. It covers setting up a TCP socket using :gen_tcp, and wrapping it in an idiomatic interface. This chapter also covers how to test a web application built using the new web server, as well as how to test the web server itself. Finally, it covers how to add concurrency to the new web server. This HTTP server will finally be used in the web framework that's being built in this book, Goldcrest.

Chapter 3, Defining Web Application Specifications Using Plug, covers the Plug package and the Plug. Conn construct, which together act as building blocks of a Phoenix HTTP request. This chapter covers the components of the Plug package and the philosophy of using it. It takes a deeper dive into Plug. Router and covers how to use it to build a routing layer to a web application. It then covers how to build a Plug adapter for a web server by taking a look at Cowboy's Plug adapter. Finally, it uses all this knowledge to build a new Plug adapter for the web server built in *Chapter 2*. This will allow us to use Plug with the new web server.

Chapter 4, Working with Controllers, leverages what we learned in *Chapter 3* about Plug and uses it to build a controller interface for the web framework, Goldcrest. It covers the basics of a controller in Phoenix and building a controller interface that follows a similar pattern. It also details how redirection works in Phoenix and how to add that functionality to the new controller interface. Finally, it uses the Plug. Test module to test the newly built controller.

Chapter 5, Adding Controller Plugs and Action Fallback, extends the controller interface built in *Chapter 4* by adding the ability to intercept a request at the controller level before letting the controller handlers handle it. This chapter covers how Phoenix handles such use cases and how we can simply use Plug to handle this. It details the Plug. Builder module that comes with the Plug package and the different approaches to test it. Finally, this chapter covers adding the ability to provide a fallback option to the controller, which handles any failed responses from all the handlers.

Chapter 6, Working with HTML and Embedded Elixir, explains how HTML is rendered on the server side by Phoenix. It introduces the EEx module, which allows us to embed Elixir between non-Elixir text, and it covers how we can use EEx to respond with dynamic server-side rendered HTML. Finally, it covers how to test the dynamically generated HTML.

Chapter 7, Working with Views, covers how to build the View interface for the web framework, Goldcrest. It goes over some of the key functionalities that can be extracted from the HTML rendering aspect of the controller and leverages the EEx module's ability to pass helper functions at the time of evaluation. Like all the other chapters, it ends by covering strategies to test the changes explained in the chapter.

Chapter 8, *Metaprogramming – Code That Writes Code*, covers metaprogramming in Elixir. It covers the constructs such as `quote`, code injection, abstract syntax trees, and so on that allow Phoenix to have such a simple interface and breaks them down in a digestible manner. It also covers macros and compile-time hooks while using these constructs to build a new **domain-specific language (DSL)** to produce music in Elixir, as an example. Finally, it covers several ways to test the meta code to make it more deterministic.

Chapter 9, *Controller and View DSL*, uses the concepts covered in the previous chapter to build a DSL around the controller and view built in *Chapter 3* to *Chapter 7*. This chapter also covers ways of making the interface easier to work with by making it easier to test and more introspective. Finally, it covers ways of testing the new `Controller` and `View` interfaces.

Chapter 10, *Building the Router DSL*, uses the metaprogramming concepts from *Chapter 8* to build a new DSL. This DSL will be for the router functionality, and this chapter covers ways to mimic Phoenix's router DSL. Like the previous chapter, it covers ways of making the DSL easier to use and test. Finally, it updates the example app built in the first part of the book to use the `Router`, `Controller`, and `View` interfaces built in the last two chapters.

To get the most out of this book

This book is very code-heavy, and in order to maximize what you learn from this book, it is recommended to code along with every code snippet. Make sure you're using the correct Elixir and Erlang versions (as recommended at the beginning of the chapters). If you are new to Elixir or haven't used it in a while, doing some practice exercises to shake off the dust before you dig into this book is also recommended.

Hardware/software covered in this book:

This book relies on Elixir 1.11.x and Erlang 23.2.x. Ensure you have `asdf` or some other package manager installed on your system. This will allow you to easily switch back and forth between Elixir and Erlang versions. This book was tested on macOS and Linux, so you may experience some inconsistencies when using Windows.

If you are using the digital version of this book, we advise you to type the code yourself or access the code from the book's GitHub repository (a link is available in the next section). Doing so will help you avoid any potential errors related to the copying and pasting of code.

Download the example code files

You can download the example code files for this book from GitHub at `https://github.com/PacktPublishing/Build-Your-Own-Web-Framework-in-Elixir`. If there's an update to the code, it will be updated in the GitHub repository.

We also have other code bundles from our rich catalog of books and videos available at https://github.com/PacktPublishing/. Check them out!

Download the color images

We also provide a PDF file that has color images of the screenshots and diagrams used in this book. You can download it here: https://packt.link/jQVwi.

Conventions used

There are a number of text conventions used throughout this book.

Code in text: Indicates code words in text, database table names, folder names, filenames, file extensions, pathnames, dummy URLs, user input, and Twitter handles. Here is an example: "In the preceding code snippet, we added a start/1 function, which listens using the default listener_options variable and creates a listening socket to accept incoming connections."

A block of code is set as follows:

```
defmodule ExperimentServer do
  # ..

  defp recv(connection_sock, messages \\ []) do
    case :gen_tcp.recv(connection_sock, 0) do
      {:ok, message} ->
        IO.puts """
        Got message: #{inspect(message)}
        """
        recv(connection_sock, [message | messages])
          {:error, :closed} ->
        IO.puts "Socket closed"
        {:ok, messages}
    end
  end
end

ExperimentServer.start(4040)
```

When we wish to draw your attention to a particular part of a code block, the relevant lines or items are set in bold.

Any command-line input or output is written as follows:

```
$ elixir experiment_server.exs
Listening on port 4040
```

Bold: Indicates a new term, an important word, or words that you see onscreen. For instance, words in menus or dialog boxes appear in **bold**.

> **Tips or important notes**
> Appear like this.

Get in touch

Feedback from our readers is always welcome.

General feedback: If you have questions about any aspect of this book, email us at customercare@packtpub.com and mention the book title in the subject of your message.

Errata: Although we have taken every care to ensure the accuracy of our content, mistakes do happen. If you have found a mistake in this book, we would be grateful if you would report this to us. Please visit www.packtpub.com/support/errata and fill in the form.

Piracy: If you come across any illegal copies of our works in any form on the internet, we would be grateful if you would provide us with the location address or website name. Please contact us at copyright@packt.com with a link to the material.

If you are interested in becoming an author: If there is a topic that you have expertise in and you are interested in either writing or contributing to a book, please visit authors.packtpub.com.

Share Your Thoughts

Once you've read, we'd love to hear your thoughts! Scan the QR code below to go straight to the Amazon review page for this book and share your feedback.

https://packt.link/r/1801812543

Your review is important to us and the tech community and will help us make sure we're delivering excellent quality content.

Download a free PDF copy of this book

Thanks for purchasing this book!

Do you like to read on the go but are unable to carry your print books everywhere?

Is your eBook purchase not compatible with the device of your choice?

Don't worry, now with every Packt book you get a DRM-free PDF version of that book at no cost.

Read anywhere, any place, on any device. Search, copy, and paste code from your favorite technical books directly into your application.

The perks don't stop there, you can get exclusive access to discounts, newsletters, and great free content in your inbox daily

Follow these simple steps to get the benefits:

1. Scan the QR code or visit the link below

https://packt.link/free-ebook/9781801812542

2. Submit your proof of purchase
3. That's it! We'll send your free PDF and other benefits to your email directly

Part 1: Web Server Fundamentals

In this part, you will learn what this book is about, the basics of web development, and how to build a web server in Elixir.

This part includes the following chapters:

- *Chapter 1, Introducing the Cowboy Web Server*
- *Chapter 2, Building an HTTP Server in Elixir*

1

Introducing the Cowboy Web Server

"Web servers are written in C, and if they're not, they're written in Java or C++, which are C derivatives, or Python or Ruby, which are implemented in C."

– Rob Pike, co-creator of Go

The web server is a key component of any modern-day web framework. Expanding on the point made in the preceding quote by Rob Pike, the **Cowboy** web server, written in **Erlang**, is also in a way implemented in C. Cowboy is the default web server used by **Phoenix**, the ubiquitous web framework in **Elixir**.

In this chapter, we will not be learning C, unfortunately, but we will take a closer look at how a web server is designed. We will provide some background on how a web server is built and set up to communicate with a client using **HyperText Markup Language** (**HTML**).

We will also learn the fundamentals of how HTTP requests and responses work, including their anatomy. We will then learn how to construct an HTTP response and send it using a web server. Moreover, we will learn the fundamentals of web server architecture by examining the components of Cowboy. Lastly, we will learn ways to test a web server and measure its performance. Doing this will put us in a better position to build our own web server in the next chapter.

The following are the topics we will cover in this chapter:

- What is a web server?
- Fundamentals of client-server architecture
- Fundamentals of HTTP
- How an HTTP server works
- Using Cowboy to build a web server

- Using dynamic routes with Cowboy

- Serving HTML

- Testing the web server

Going through these topics and looking at Cowboy will allow us to build our own HTTP server in *Chapter 2*.

Technical requirements

The best way to work through this chapter is by following along with the code on your computer. So, having a computer with Elixir and Erlang ready to go would be ideal. I recommend using a version manager such as `asdf` to install *Elixir 1.11.x* and *Erlang 23.2.x*, to get similar results as the code written in the book. We will also be using an HTTP client such as cURL or Wget to make HTTP requests to our server, and a web browser to render HTML responses.

Although most of the code in this chapter is relatively simple, basic knowledge of Elixir and/or Erlang would also come in handy. It will allow you to get more out of this chapter while setting the foundation for other chapters.

Since most of this chapter isn't coding, you can also choose to read without coding, but the same doesn't apply to other chapters.

The code examples for this chapter can be found at `https://github.com/PacktPublishing/Build-Your-Own-Web-Framework-in-Elixir/tree/main/chapter_01`

What is a web server?

A web server is an entity that delivers the content of a site to the end user. A web server is typically a long-running process, listening for requests on a port, and upon receiving a request, the web server responds with a document. This way of communication is standardized by the **Transmission Control Protocol/Internet Protocol** (**TCP/IP**) model, which provides a set of communication protocols used by any communication network. There are other layers of standardization, such as the **HyperText Transfer Protocol** (**HTTP**) and **File Transfer Protocol** (**FTP**), which dictate standards of communication at the application layer based on different applications such as web applications in the case of HTTP, and file transfer applications in the case of FTP, while still using TCP/IP at the network layer. In this book, we will be primarily focusing on a web server using HTTP at the application layer.

Example HTTP server

If you have Python 3 installed on your machine (you likely do), you can quickly spin up a web server that serves a static HTML document by creating an `index.html` file in a new directory and running a simple HTTP Python server. Here are the commands:

```
$ mkdir test-server && cd test-server && touch index.html
$ echo "<h1>Hello World</h1>" > index.html
$ python -m http.server 8080
Serving HTTP on 0.0.0.0 port 8080 (http://0.0.0.0:8080/) . . .
```

If you are on Python 2, replace `http.server` with `SimpleHTTPServer`.

Now, once you navigate to `http://localhost:8080/` on your web browser, you should see `"Hello World"` as the response. You should also be able to see the server logs when you navigate back to the terminal.

To stop the HTTP server, press *Ctrl + C*.

The primary goal of web servers is to respond to a client's request with documents in the form of HTML or JSON. These days, however, web servers do much more than that. Some have analytical features, such as an Admin UI, and some have the ability to generate dynamic documents. For example, Phoenix's web server has both of those features. Now that we know what a web server is, let's learn about how it is used with the client-server architecture.

Exploring the client-server architecture

In the context of HTTP servers, **clients** generally mean the web browsers that enable end users to read the information being served, whereas **servers** mean long-running processes that serve information in the form of documents to those clients. These documents are most commonly written in **HTML** and are used as a means of communication between the client and the server. Clients are responsible for enabling the end user to send a request to the server and display the response from the server. Browsers allow the users to retrieve and display information without requiring any knowledge of HTML or web servers, by just providing an address (the URL).

At a given time, many clients can access a server's information. This puts the burden of scaling on the servers as they need to be designed with the ability to respond to multiple requests within an acceptable period of time. Now that we understand a web server's primary goal, let's move on to the protocol that enables communication between web servers: **HTTP**.

Understanding HTTP

HTTP is an application layer protocol that provides communication standards between clients (such as web browsers) and web servers. This standardization helps browsers and servers talk to each other as long as the request and the response follow a specific format.

An HTTP request is a text document with four elements:

- **Request line:** This line contains the HTTP method, the resource requested (URI), and the HTTP version being used for the request. The HTTP method generally symbolizes the intended action being performed on the requested resource. For example, GET is used to retrieve resource information, whereas POST is used to send new resource information as a form.

- **Request headers:** The next group of lines contains headers, which contain information about the request and the client. These headers are usually used for authorization, determining the type of request or resource, storing web browser information, and so on.

- **Line break:** This indicates the end of the request headers.

- **Request body (optional):** If present, this information contains data to be passed to the server. This is generally used to submit domain-specific forms.

Here's an example of an HTTP request document:

```
GET / HTTP/1.1
Host: localhost:8080
User-Agent: curl/7.75.0
Accept: */*

Body of the request
```

As you can see, the preceding request was made with the GET method to localhost:8080 with the body, Body of the request.

Similarly, an HTTP response contains four elements:

- **Response status line:** This line consists of the HTTP version, a status code, and a reason phrase, which generally corresponds to the status code. Status codes are three-digit integers that help us categorize responses. For example, 2XX status codes are used for a successful response, whereas 4XX status codes are used for errors due to the request.

- **Response headers:** Just like request headers, this group of lines contains information about the response and the server's state. These headers are usually used to show the size of the response body, server type, date/time of the response, and so on.

- **Line break:** This indicates the end of response headers.

- **Response body (optional):** If present, this section contains the information being relayed to the client.

The following is an example of an HTTP response document:

```
HTTP/1.1 404 Not Found
content-length: 13
content-type: text/html
server: Cowboy

404 Not found
```

The preceding response is an example of a 404 (Not found) response. Notice that content-length shows the number of characters present in the response body.

Now that we know how HTTP facilitates client-server communication, it is time to build a web server using Cowboy.

Understanding Cowboy's architecture

Cowboy is a minimal and fast HTTP web server written in Erlang. It supports several modern standards, such as HTTP/2, HTTP/1.1, and WebSocket, for example. On top of that, it also has several introspective capabilities, thus enabling easier development and debugging. Cowboy has a very well-written and well-documented code base, with a highly extendable design, which is why it is the default web server for the Phoenix framework.

Cowboy uses **Ranch**, a TCP socket accepter, to create a new TCP connection, on top of which it uses its router to match a request to a handler. Routers and handlers are middleware that are part of Cowboy. Upon receiving a request, Cowboy creates a stream, which is further handled by a stream handler. Cowboy has a built-in configuration that handles a stream of requests using :cowboy_stream_h. This module spins up a new Erlang process for every request that is made to the router.

Cowboy also sets up one process per TCP connection. This also allows Cowboy to be compliant with HTTP/2, which requires concurrent requests. Once a request is served, the Erlang process is killed without any need for cleanup.

The following figure shows the Cowboy request/response cycle:

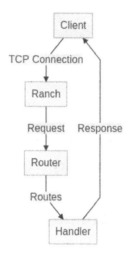

Figure 1.1 – Cowboy request/response cycle

As you can see in *Figure 1.1*, when a client makes a request, Ranch first converts it into a stream, which is further handled by the router and handler middleware in Cowboy. Traditionally, a response is sent either by the router or the handler. For example, a handler could handle a request and send a response or, if no handler is present for a route, the router could also send a 404 response.

Cowboy also generates a few response headers, as we will see in the next section, where we build and test a Cowboy-powered web application.

Building a web application using Cowboy

In this section, we will take a look at some of the individual components of the Cowboy web server and use them to build a simple web application in Elixir.

Creating a new Mix project

Let's start by creating a new Mix project by entering the following in your terminal:

```
$ mix new cowboy_example --sup
```

> **What is Mix?**
>
> Mix is a build tool written in Elixir. Its purpose is to bundle all the dependencies required by a project and provide an interface to run tasks that rely on the application environment. If you're familiar with the Ruby world, you can think of Mix as a combination of Rake and Bundler.

Passing the ‑‑sup option to the mix new command allows us to create a Mix project with a **supervision tree**. A supervision tree (or a supervisor) is a process that simply monitors other processes and is responsible for automatically restarting any process within the tree if it fails or crashes. We will be using the supervision tree in this application to start and supervise our web server process to make sure it is started when the application is started and to ensure that it keeps running.

Now, we will add Cowboy as a dependency to our project by adding it to the mix.exs file's dependencies:

mix.exs

```
defmodule CowboyExample.MixProject do
  # ...
  defp deps do
    [
      {:cowboy, "~> 2.8"}
    ]
  end
end
```

Follow it up by running mix deps.get from the project's root directory, which should fetch Cowboy as a dependency.

Adding a handler and router to Cowboy

Now that we have added Cowboy, it is time to configure it to listen on a port. We will be using two functions to accomplish that:

- :cowboy_router.compile/1: This function is responsible for defining a set of requested hosts, paths, and parameters to a dedicated request handler. This function also generates a set of rules, known as **dispatch rules**, to use those handlers.

- :cowboy.start_clear/3: This function is responsible for starting a listener process on a TCP channel. It takes a listener name (an atom), transport options such as the TCP port, and protocol options such as the dispatch rules generated using the :cowboy_router. compile/1 function.

Now, let us use these functions to write an HTTP server. We can start by creating a new module to house our new HTTP server:

lib/cowboy_example/server.ex

```
defmodule CowboyExample.Server do
  @moduledoc """
  This module defines a cowboy HTTP server and starts it
  on a port
```

```
  """

  @doc """
  This function starts a Cowboy server on the given port.

  Routes for the server are defined in CowboyExample.Router
  """
  def start(port) do
    routes = CowboyExample.Router.routes()

    dispatch_rules =
      :cowboy_router.compile(routes)

    {:ok, _pid} =
      :cowboy.start_clear(
        :listener,
        [{:port, port}],
        %{env: %{dispatch: dispatch_rules}}
      )
  end
end
```

The preceding function starts a Cowboy server that listens to HTTP requests at the given port. By pattern matching on {:ok, _pid}, we're making sure that :cowboy.start_clear/3 doesn't fail silently.

We can try to start our web server by calling the CowboyExample.Server.start/1 function with a port. But, as you can see, we will also need to define the CowboyExample.Router router module. This module's responsibility is to define routes that can be used to generate dispatch rules for our HTTP server. This can be done by storing all the routes, parameters, and handler tuples in the router module and passing them to the :cowboy_router.compile/1 call.

Let's define the router module with the route for the root URL of the host (/):

lib/cowboy_example/router.ex

```
defmodule CowboyExample.Router do
  @moduledoc """
  This module defines all the routes, params and handlers.

  This module is also the handler module for the root
  route.
  """

  require Logger
```

```
@doc """
Returns the list of routes configured by this web server
"""
def routes do
  [
    # For now, this module itself will handle root
    # requests
    {:_, [{"/", __MODULE__, []}]}
  ]
end
end
```

We will also be using `CowboyExample.Router` itself as the handler for that route, which means we have to define the `init/2` function, which takes the request and its initial state.

So, let us define the `init/2` function:

lib/cowboy_example/router.ex

```
defmodule CowboyExample.Router do
  # ..
  @doc """
  This function handles the root route, logs the requests
  and responds with Hello World as the body
  """
  def init(req0, state) do
    Logger.info("Received request: #{inspect req0}")

    req1 =
      :cowboy_req.reply(
        200,
        %{"content-type" => "text/html"},
        "Hello World",
        req0
      )

    {:ok, req1, state}
  end
end
```

As you can see in the preceding code, we have defined the `routes/0` function, which returns the dispatch rules routes for our web application. For the handler module, we're currently using the `CowboyExample.Router` module itself by defining the `init/2` function, which responds with `"Hello World"` and a status of `200`, whenever invoked. We have also added a call to the `Logger` module to log all requests to the handler. This will increase visibility while running it in the development environment.

In order for our web server to start up when the app is started, we need to add it to our application's supervision tree.

Supervising the web server

Now that we have added a router and a handler to our web server, we can add it as a child to our supervision tree by updating the list of children in our application module. For now, I will use a hardcoded TCP port of `4040` for our server, but we will use application-level configurations to set it later in this chapter:

lib/cowboy_example/application.ex

```
defmodule CowboyExample.Application do
  @moduledoc false

  use Application

  @impl true
  def start(_type, _args) do
    children = [
      # Add this line
      {Task, fn -> CowboyExample.Server.start(4040) end}
    ]

    opts = [
      strategy: :one_on_one,
      name: CowboyExample.Supervisor
    ]

    Supervisor.start_link(children, opts)
  end
end
```

In the preceding code, we're adding to the supervised children a `Task` with the function to start the Cowboy listener as an argument that eventually gets passed to `Task.start_link/1`. This makes sure that our web server process is part of the application's supervision tree.

Now, we can run our web application by running the `mix` project with the `--no-halt` option:

```
$ mix run --no-halt
```

> **Note**
>
> Passing the `--no-halt` option to the `mix run` command makes sure that the application, along with the supervision tree, is still running even after the command has returned. This is generally used for long-running processes such as web servers.

Without stopping the previous command, in a separate terminal session, we can make a request to our web server using the cURL command-line utility with the −v option to get a verbose description of our requests and responses:

```
$ curl -v http://localhost:4040/
*     Trying ::1:4040...
* connect to ::1 port 4040 failed: Connection refused
*     Trying 127.0.0.1:4040...
* Connected to localhost (127.0.0.1) port 4040 (#0)
> GET / HTTP/1.1
> Host: localhost:4040
> User-Agent: curl/7.75.0
> Accept: */*
>
* Mark bundle as not supporting multiuse
< HTTP/1.1 200 OK
< content-length: 11
< content-type: text/html
< server: Cowboy
<
* Connection #0 to host localhost left intact
Hello world%
```

As we can see in the preceding code, we get the expected "Hello World" response along with the expected status code of 200. As mentioned in the previous section, Cowboy adds custom response headers to give us more information about how it was processed. We can also see headers for the type of server (Cowboy), content length, and content type.

We should also see an application-level log corresponding to the request in the terminal session running the `mix` project. The logs should look somewhat like this:

```
$ mix run --no-halt

20:39:43.061 [info]   Received request: %{
  bindings: %{},
```

```
    body_length: 0,
    cert: :undefined,
    has_body: false,
    headers: %{
      "accept" => "*/*",
      "host" => "localhost:4040",
      "user-agent" => "curl/7.75.0"
    },
    host: "localhost",
    host_info: :undefined,
    method: "GET",
    path: "/",
    path_info: :undefined,
    peer: {{127, 0, 0, 1}, 35260},
    pid: #PID<0.271.0>,
    port: 4040,
    qs: "",
    ref: :listener,
    scheme: "http",
    sock: {{127, 0, 0, 1}, 4040},
    streamid: 1,
    version: :"HTTP/1.1"
}
```

We can see that we're logging all the details corresponding to the request including headers, the host, the URI, and the process ID of the process processing the request.

Congratulations, you have now successfully built a `Hello World` web server using Cowboy. Now, it's time to add more routes to our web server.

Adding routes with bindings

Most web applications support the ability to serve not only a static route but also dynamic routes with a specific pattern. It's time to see how we can leverage Cowboy to add dynamic routes to our router.

Say we want to add a new route to our application that responds with a custom greeting for a person whose name is dynamic. Let's update our router to define a handler for a new dynamic route. We can also use this opportunity to move our `Root` handler (the `init/2` function) to a different module. This makes our code more compliant with the single-responsibility principle, making it easier to follow:

lib/cowboy_example/router.exdefmodule

```
CowboyExample.Router do
  @moduledoc """
  This module defines all the routes, params and handlers.
```

```
    """
    alias CowboyExample.Router.Handlers.{Root, Greet}

    @doc """
    Returns the list of routes configured by this web server
    """
    def routes do
      [
        {:_, [
          {"/", Root, []},
          # Add this line
          {"/greet/:who", [who: :nonempty], Greet, []}
        ]}
      ]
    end
  end
```

In the preceding code, we have added a new route that expects a non-empty value for the :who variable binding. This variable gets bound to a request based on the URL. For example, for a request with the URL "/greet/Luffy", the variable bound to :who will be "Luffy", and for a request with the URL "/greet/Zoro", it will be "Zoro".

Now, let's define the Root handler and move the init/2 function from our router to the new handler module. This separates the concerns of defining routes and handling requests:

lib/cowboy_example/router/handlers/root.ex

```
defmodule CowboyExample.Router.Handlers.Root do
  @moduledoc """
  This module defines the handler for the root route.
  """

  require Logger

  @doc """
  This function handles the root route, logs the requests
  and responds with Hello World as the body
  """
  def init(req0, state) do
    Logger.info("Received request: #{inspect req0}")

    req1 =
      :cowboy_req.reply(
        200,
        %{"content-type" => "text/html"},
        "Hello World",
```

```
        req0
      )

    {:ok, req1, state}
  end
end
```

Similarly, let's define the `Greet` handler for our new dynamic route. We know that the request has a variable binding corresponding to the : who key by the time it gets to this handler. Therefore, we can use the : cowboy_req.binding/2 function to access the value of : who bound to the request:

lib/cowboy_example/router/handlers/greet.ex

```elixir
defmodule CowboyExample.Router.Handlers.Greet do
  @moduledoc """
  This module defines the handler for "/greet/:who" route.
  """

  require Logger

  @doc """
  This function handles the "/greet/:who", logs the
  requests and responds with Hello `who` as the body
  """
  def init(req0, state) do
    Logger.info("Received request: #{inspect req0}")

    who = :cowboy_req.binding(:who, req0)

    req1 =
      :cowboy_req.reply(
        200,
        %{"content-type" => "text/html"},
        "Hello #{who}",
        req0
      )

    {:ok, req1, state}
  end
end
```

In the preceding code snippet, we get the value bound to : who for the request and use it with string interpolation to call "Hello : who". Now, we have two valid routes for our web server: the root and the dynamic greet route.

We can test our updates by restarting the Mix application. That can be done by stopping the HTTP server using *Ctrl + C*, followed by running `mix run --no-halt` again. Now, let's make a request to test the new route with `Elixir` as `:who`:

```
$ curl http://localhost:4040/greet/Elixir
Hello Elixir%
```

Cowboy offers another way to add dynamic behavior to our routes, and that is by passing query parameters to our URL. Query parameters can be captured by using the `:cowboy_req.parse_qs/1` function. This function takes a binding name (`:who` in this case) and the request itself. Let's update our `greet` handler to now take a custom query parameter for `greeting` that overrides the default `"Hello"` greeting, which we can put in a module attribute for better code organization:

lib/cowboy_example/router/handlers/greet.ex

```elixir
defmodule CowboyExample.Router.Handlers.Greet do
  # ..
  @default_greeting "Hello"
  # ..
  def init(req0, state) do
    greeting =

    # ..
      req0
      |> :cowboy_req.parse_qs()
      |> Enum.into(%{})
      |> Map.get("greeting", @default_greeting)

    req1 =
      :cowboy_req.reply(
        200,
        %{"content-type" => "text/html"},
        "#{greeting} #{who}",
        req0
      )

    {:ok, req1, state}
  end
end
```

We have now updated our greet handler to use :cowboy.parse_qs/1 to fetch query parameters from the request. We then put those matched parameters into a map and get the value in the map corresponding to the "greeting" key, with a default of "Hello". Now, the greet route should take a "greeting" query parameter to update the greeting used to greet someone in the response. We can test our updates again by restarting the application and making a request to test the route with a custom greeting parameter:

```
$ curl http://localhost:4040/greet/Elixir\?greeting=Hola
Hola Elixir%
```

We now have a web server with a fully functional dynamic route. You might have noticed that we didn't specify any HTTP method while defining the routes. Let us see what happens when we try to make a request to the root with the POST method:

```
$ curl -X POST http://localhost:4040/
Hello World%
```

As you can see in the example, our web server still responded to the POST request in the same manner as GET. We don't want that behavior so, in the next section, we will see how to validate the HTTP method of a request and restrict the root of our application to only respond to GET requests.

Validating HTTP methods

Most modern web applications have a way of restricting requests to a route based on the HTTP method. In this section, we will see how to restrict our handlers to work with a specific HTTP method in a Cowboy-based web application. The simplest way of accomplishing that in a Cowboy handler is by pattern matching on the first argument of the init/2 function, which is the request. A Cowboy request contains a lot of information, including the HTTP method used to make the request. So, by pattern matching on the request with a specific HTTP method, we can filter requests based on HTTP methods. However, we will also be needing a general clause for the init/2 function, which responds with a 404 error.

In the Greet handler, let us update init/2 to match only on requests with the GET method, along with another clause that responds with a 404 (Not Found) error:

lib/cowboy_example/router/handlers/greet.ex

```
defmodule CowboyExample.Router.Handlers.Greet do
  # ..
  def init(%{method: "GET"} = req0, state) do
  # ..
  end

  # General clause for init/2 which responds with 404
```

```
def init(req0, state) do
  Logger.info("Received request: #{inspect req0}")

  req1 =
    :cowboy_req.reply(
      404,
      %{"content-type" => "text/html"},
      "404 Not found",
      req0
    )

  {:ok, req1, state}
  end
end
```

Now, let's make sure only GET requests are accepted by our server for the route. Let's first make sure GET requests are working:

```
$ curl http://localhost:4040/greet/Elixir\?greeting=Hola
Hola Elixir%
```

It's time to check that a POST request for the `greet` route returns a `404` error:

```
$ curl http://localhost:4040/greet/Elixir\?greeting=Hola -X POST
404 Not found%
```

This ensures that our route works only for GET requests. Another way of validating HTTP methods of our requests would be by using Cowboy middleware, but we will not be covering that in this chapter.

Cowboy middleware

In Cowboy, middleware is a way to process an incoming request. Every request has to go through two types of middleware (the router and the handler), but Cowboy allows us to define our own custom middleware module, which gets executed in the given order. A custom middleware module just needs to implement the `execute/2` callback defined in the `cowboy_middleware` behavior.

Great! We have successfully enforced a method type for a route. Next, we will learn how to serve HTML files instead of raw strings.

Responding with HTML files

Generally, when we write web servers, we do not write our HTML as strings in handlers. We write our HTML in separate files that are served by our server. We will use our application's `priv` directory to store these static files. So, let's create a `priv/static` folder in the root of our project and add an `index.html` file in that folder. To add some HTML, we can use this command:

```
$ echo "<h1>Hello World</h1>" > priv/static/index.html
```

> **The priv directory in OTP**
>
> In OTP (Open Telecom Platform or Erlang) and Elixir, the `priv` directory is a directory specific to an application that is intended to store files needed by the application when it is running. Phoenix, for example, uses the `priv/static` directory to store processed JavaScript and CSS assets for runtime usage.

Let's add an endpoint to our server that returns a static HTML file:

lib/cowboy_example/router.ex

```elixir
defmodule CowboyExample.Router do
  @moduledoc """
  This module defines routes and handlers for the web
  server
  """

  alias CowboyExample.Router.Handlers.{Root, Greet, Static}

  @doc """
  Returns the list of routes configured by this web server
  """
  def routes do
    [
      {:_, [
        {"/", Root, []},
        {"/greet/:who", [who: :nonempty], Greet, []},
        # Add this line
        {"/static/:page", [page: :nonempty], Static, []}
      ]}
    ]
  end
end
```

Now, we need a static handler module, which will look for and respond with the given page in the /priv/static folder and, if not found, will return a 404 error:

lib/cowboy_example/router/handlers/static.ex

```elixir
defmodule CowboyExample.Router.Handlers.Static do
  @moduledoc """
  This module defines the handler for "/static/:page"
  route.
  """
  require Logger

  @doc """
  This handles "/static/:page" route, logs the requests and
  responds with the requested static HTML page.

  Responds with 404 if the page isn't found in the
  priv/static folder.
  """
  def init(req0, state) do
    Logger.info("Received request: #{inspect req0}")

    page = :cowboy_req.binding(:page, req0)

    req1 =
      case html_for(page) do
        {:ok, static_html} ->
          :cowboy_req.reply(
            200,
            %{"content-type" => "text/html"},
            static_html,
            req0
          )

        _ ->
          :cowboy_req.reply(
            404,
            %{"content-type" => "text/html"},
            "404 Not found",
            req0
          )
      end

    {:ok, req1, state}
  end
```

```
defp html_for(page) do
  priv_dir =
    :cowboy_example
    |> :code.priv_dir()
    |> to_string()

  page_path = priv_dir <> "/static/#{page}"

  File.read(page_path)
  end
end
```

In the preceding module, the `html_for/1` function is responsible for fetching the HTML files from our application's `priv` directory, for a given path. If the file is present, the function returns `{:ok, <file_contents>}, >}`; otherwise, it returns an error, upon which we will respond with a `404` message.

We can test the preceding route by restarting our server again and making a request to the `/static/index.html` path. But this time, let us use the web browser in order to render the HTML contents properly. Here's what you should see:

Hello World

Figure 1.2 – Successful HTML response

Also, to make sure our `404` handler is working correctly, let's make a browser request to `/static/bad.html`, a file not present in our application's `priv` directory. You should see a `404` message:

404 Not found

Figure 1.3 – Failed HTML response

Now, we have a web server that can respond with static HTML files. It's time to see how we can go about testing it.

Testing the web server with ExUnit

Automated testing is a key part of any software, especially in a dynamic-typed language such as Elixir. It is one of the catalysts for writing deterministic software while documenting the expected behaviors of its components. Due to this reason, we will be making an effort to test everything we build in this book, including the Cowboy-powered web application we have built in this chapter.

In order to test our web application, we first need to be able to run our application on a different port in the test environment. This is to ensure that other /static/bad.html environments do not interfere with our tests. We also can use an application-level configuration to set a port on which the Cowboy server listens to all the requests. This will allow us to separate the test port from the development port.

So, let's update our application to use the configured port or default it to 4040 using an @port module attribute:

lib/cowboy_example/application.ex

```
defmodule CowboyExample.Application do
  @moduledoc false

  use Application

  @port Application.compile_env(
          :cowboy_example,
          :port,
          4040
        )

  @impl true
  def start(_type, _args) do
    children = [
      # Add this line
      {Task, fn -> CowboyExample.Server.start(@port) end}
    ]

    opts = [
      strategy: :one_for_one,
      name: CowboyExample.Supervisor
    ]

    Supervisor.start_link(children, opts)
  end
end
```

We can make sure that the application configuration is different for different Mix environments by adding the config/config.exs file, and setting a different port in our config for the test environment. We will also be updating the logger to not log warnings. So, let's add a config file with the following contents:

config/config.exs

```
import Config

if Mix.env() == :test do

  config :cowboy_example,
    port: 4041

  config :logger, warn: false
end
```

> **Note**
> Mix.Config has been deprecated in newer versions of Elixir. You might have to use the Config module instead.

Now, let's add tests for our server endpoints. In order to test our web server, we need to make HTTP requests to it and test the responses. To make HTTP requests in Elixir, we will be using **Finch**, a lightweight and high-performance HTTP client written in Elixir.

So, let's add Finch to our list of dependencies:

mix.exs

```
defmodule CowboyExample.MixProject do
  # ...
  defp deps do
    [
      {:cowboy, "~> 2.8"},
      {:finch, "~> 0.6"}
    ]
  end
end
```

Running mix deps.get will fetch Finch and all its dependencies.

Now, let's add a test file to test our server. In the test file, we will be setting up Finch to make HTTP calls to our server. In this section, we will only be testing the happy paths (200 responses) of our root and greet endpoints:

test/cowboy_example/server_test.exs

```elixir
defmodule CowboyExample.ServerTest do
  use ExUnit.Case

  setup_all do
    Finch.start_link(name: CowboyExample.Finch)

    :ok
  end

  describe "GET /" do
    test "returns Hello World with 200" do
      {:ok, response} =
        :get
        |> Finch.build("http://localhost:4041")
        |> Finch.request(CowboyExample.Finch)

      assert response.body == "Hello World"
      assert response.status == 200
      assert {"content-type", "text/html"} in response.headers
    end
  end

  describe "GET /greeting/:who" do
    test "returns Hello `:who` with 200" do
      {:ok, response} =
        :get
        |> Finch.build("http://localhost:4041/greet/Elixir")
        |> Finch.request(CowboyExample.Finch)

      assert response.body == "Hello Elixir"
      assert response.status == 200
      assert {"content-type", "text/html"} in response.headers
    end

    test "returns `greeting` `:who` with 200" do
      {:ok, response} =
        :get
        |> Finch.build("http://localhost:4041/greet/
                       Elixir?greeting=Hola")
```

```
        |> Finch.request(CowboyExample.Finch)

      assert response.body == "Hola Elixir"
      assert response.status == 200
      assert {"content-type", "text/html"} in response.headers
    end
  end
end
```

As you can see in the preceding module, we have added tests for the two endpoints using Finch. We make calls to our server, running on port 4041 in the test environment, with different request paths and parameters. We then test the response's body, status, and headers.

This should give you a good idea of how to go about testing a web server. Over the next few chapters, we will be building on top of this foundation and coming up with better ways of testing our web server.

Summary

When I first learned about how a web server works, I was overwhelmed with the number of things that go into building one. Then I decided to look at the code of Puma, a web server written in Ruby, which is also used by Rails. I was surprised by how much more I learned by just looking into Puma than by reading articles about web servers. It is due to that reason that we are kicking off this book by looking at Cowboy. I believe that learning about the basics of Cowboy will better position us to build our own web server in the next few chapters.

In this chapter, we first learned the basics of a web server along with the client-server architecture. We also looked at the high-level architecture of Cowboy and learned about how some of its components such as the router and handlers work. We also added dynamic behavior to our routes by using path variables and query parameters, followed by serving static HTML files. We finished by learning how to test our routes using an HTTP client. In the next chapter, we will use what we learned in this chapter to build our own HTTP server from scratch.

Exercises

Some of you might have noticed that we haven't tested a few aspects of our web server. Using what you have learned in this chapter, complete these exercises:

- Write test cases for our web server that would lead to 404 responses
- Write tests for the static route that respond with HTML files

There are other (better) ways of testing an HTML response, which we haven't covered in this chapter. We will dig deeper into those testing methods later in this book.

2

Building an HTTP Server in Elixir

After learning the basics of a web server and taking a deeper look into Cowboy in the previous chapter, in this chapter, we will see how to build an HTTP server in Elixir. In order to build the HTTP server, we will leverage Erlang's `:gen_tcp` module. Using that module, we will create a TCP/IP connection between the client and the server, just like in Cowboy. We will then write a configurable web server package that will allow other apps to provide their own handlers to respond with. We will also be testing our HTTP server thoroughly using unit tests and a helper application. We will close up by making our HTTP server concurrent.

After reading through this chapter, you will be able to build an HTTP server from scratch using Elixir and have a better understanding of how TCP works to facilitate communication using sockets. We will be covering the following topics:

- Listening over a TCP socket using `:gen_tcp`
- Responding over a TCP socket using `:gen_tcp`
- Writing a web server package in Elixir
- Leveraging application configuration to make the web server flexible
- Testing the newly written web server package using unit tests
- Testing the web server package using another application
- Adding concurrency to the web server

Technical requirements

This chapter is very code-heavy since we will be implementing most of Cowboy's features that we learned about in *Chapter 1*. In order to get the most out of this chapter, it is recommended to code along with every snippet and use Elixir 1.11.x and Erlang 23.2.x to get similar results. We will also be using the cURL and Wget command-line utilities, so having both of those ready on your machine will be useful.

As far as technical knowledge goes, I have tried to keep this chapter self-sufficient. However, basic knowledge of Elixir, Erlang, supervision trees, and TCP/IP will help you understand this chapter faster. In order to better understand some of the technical terms explained in this chapter, please read this chapter after *Chapter 1*.

The code examples for this chapter can be found at `https://github.com/PacktPublishing/Build-Your-Own-Web-Framework-in-Elixir/tree/main/chapter_02`

Listening over a TCP socket using :gen_tcp

Cowboy uses Ranch to create and manage a TCP connection, but in our web server, we will be using the `:gen_tcp` module. This module allows two entities to communicate using TCP/IP sockets and is shipped as a part of Erlang's **Open Telecom Platform** (**OTP**).

We need to first listen on a TCP/IP port and accept requests on a TCP socket. Once the request is processed and a response is sent, we also need to close the connection and socket.

Here are the functions we will be using:

- `:gen_tcp.listen/2`: Accepts a listening port and server options. Creates a listening socket that listens on a given TCP/IP port.

- `:gen_tcp.accept/2`: Accepts a listening socket returned by `:gen_tcp.listen/2` and a timeout, which defaults to infinity. This function creates another socket that represents a connection between the server and the client. This socket is used to send a response to the client from the server. Since web servers are long-running processes in Elixir, we will be using the default timeout of infinity in this chapter.

- `:gen_tcp.recv/2`: Accepts a socket, length, and timeout. The socket here is the socket used to respond to the client, which is the same as the socket returned by `:gen_tcp.accept/2`. Length corresponds to the length of the bytes returned. We will be using 0 as our length, which implies that all the bytes are returned. The timeout here also defaults to infinity, which we will be using for our web server. This function responds with a packet tuple that contains information corresponding to the request that can further be used for route-matching and method validations.

Let us start by creating a .exs file to run our experiments in a separate folder:

```
$ touch experiment_server.exs
```

This is where we will write our server using :gen_tcp and eventually extract relevant code into a web server module that will be packaged with *Goldcrest*, our web framework.

Now, let's define an ExperimentServer module with a function, start/1, which starts a TCP connection socket on a given TCP port without a response:

---

experiment_server.exs

```elixir
defmodule ExperimentServer do
  def start(port) do
    listener_options = [active: false, packet: :http_bin]

    {:ok, listen_socket} =
     :gen_tcp.listen(
       port,
       listener_options
     )

    IO.puts "Listening on port #{port}"

    {:ok, connection_sock} = :gen_tcp.accept(listen_socket)

    {:ok, messages} = recv(connection_sock)
    :ok = :gen_tcp.close(connection_sock)
    :ok = :gen_tcp.close(listen_socket)

    IO.puts """
    Messages:
    #{inspect(messages)}
    """

    :ok
  end
end
```

In the preceding code snippet, we added a start/1 function that listens using the default listener_ options variable and creates a listening socket to accept incoming connections. It then calls a private function, recv/1, which handles incoming requests. Finally, it closes the connection and the listening socket.

Let's now implement the `recv/1` function:

experiment_server.exs

```elixir
defmodule ExperimentServer do
  # ..

  defp recv(connection_sock, messages \\ []) do
    case :gen_tcp.recv(connection_sock, 0) do
      {:ok, message} ->
        IO.puts """
        Got message: #{inspect(message)}
        """
        recv(connection_sock, [message | messages])
          {:error, :closed} ->
        IO.puts "Socket closed"
        {:ok, messages}
    end
  end
end

ExperimentServer.start(4040)
```

To create a listening socket, we used `:gen_tcp.listen/2` with `[active: false, packet: http_bin]` listener options. Let's have a look at this in more detail:

- `active: false`: Means packets will not be delivered as messages but will be retrieved by calling `:gen_tcp.recv/2`

- `packet: :http_bin`: Sets the expected format of the packets as an HTTP binary packet, which will further be used by `:erlang.decode_packet/3` to decode the incoming packet

We then passed the listening socket to the `:gen_tcp.accept/2` function, which created a connection socket that we can further use with `:gen_tcp.recv/2` to receive a message on the connection socket. We also added a `receive` loop to keep receiving messages on that connection socket until the connection is closed by the client. We did that by writing the `recv/2` recursive function.

Once the socket is closed (for example, by receiving a *Ctrl + C* from the client side), all the messages received on that connection are printed and both the listening socket and the connection socket are closed.

Finally, to start the process, we call `ExperimentServer.start/1` on port `4040`.

We can now run the preceding `.exs` file to start the TCP listener process:

```
$ elixir experiment_server.exs
Listening on port 4040
```

Now, without stopping the process, let's send HTTP calls to that port using `curl`:

```
$ curl -v http://localhost:4040/
*   Trying ::1:4040...
* connect to ::1 port 4040 failed: Connection refused
*   Trying 127.0.0.1:4040...
* Connected to localhost (127.0.0.1) port 4040 (#0)
> GET / HTTP/1.1
> Host: localhost:4040
> User-Agent: curl/7.75.0
> Accept: */*
>
```

You will see logs in the session running our TCP process as well:

```
Listening on port 4040

Got message:
  {:http_request, :GET, {:abs_path, "/"}, {1, 1}}

Got message:
  {:http_header, 14, :Host, "Host", "localhost:4040"}

Got message:
  {:http_header, 24, :"User-Agent", "User-Agent",
  "curl/7.75.0"}

Got message:
  {:http_header, 8, :Accept, "Accept", "*/*"}

Got message: :http_eoh
```

To close the socket, hit *Ctrl + C* from the session running the `curl` request.

This will close the connection socket and print the messages as received on the connection, as follows:

```
$ elixir experiment_server.exs
Listening on port 4040

Got message:
  {:http_request, :GET, {:abs_path, "/"}, {1, 1}}
```

```
Got message:
  {:http_header, 14, :Host, "Host", "localhost:4040"}

Got message:
  {:http_header, 24, :"User-Agent", "User-Agent",
   "curl/7.75.0"}

Got message:
  {:http_header, 8, :Accept, "Accept", "*/*"}

Got message: :http_eoh

Socket closed

Messages:

[
  :http_eoh,
  {:http_header, 8, :Accept, "Accept", "*/*"},
  {:http_header, 24, :"User-Agent", "User-Agent",
   "curl/7.75.0"},
  {:http_header, 14, :Host, "Host", "localhost:4040"},
  {:http_request, :GET, {:abs_path, "/"}, {1, 1}}
]
```

We now know how to use `:gen_tcp` to establish a TCP connection. Next, we will update the `ExperimentServer` module to respond to incoming requests.

Responding over a TCP socket using :gen_tcp

Now that we know how to use `:gen_tcp` to initiate a TCP/IP connection, the next step is to respond using that connection. In order to respond to a request, we will be using the `:gen_tcp.send/2` function.

The `:gen_tcp.send/2` function allows us to send I/O data (generally string) on a connection socket. It doesn't support send timeout, so it only takes a connection socket and data that needs to be sent. We can use the connection socket returned by the `:gen_tcp.accept/1` call in the previous section to send an HTTP response.

In order to send data over a connection socket, let's restructure our `ExperimentServer` module:

experiment_server.exs

```
defmodule ExperimentServer do
  require Logger
```

```elixir
  def start(port) do
    listener_options = [
    active: false,
    packet: :http_bin,
    reuseaddr: true
  ]

    {:ok, listen_socket} =
    :gen_tcp.listen(
      port,
      listener_options
    )

    Logger.info("Listening on port #{port}")

    listen(listen_socket)

    :gen_tcp.close(listen_socket)
  end

  defp listen(listen_socket) do
    {:ok, connection_sock} = :gen_tcp.accept(listen_socket)
    {:ok, req} = :gen_tcp.recv(connection_sock, 0)

    Logger.info("Got request: #{inspect req}")
    respond(connection_sock)
    listen(listen_socket)
  end

  defp respond(connection_sock) do
    :gen_tcp.send(connection_sock, "Hello World")

    Logger.info("Sent response")

    :gen_tcp.close(connection_sock)
  end
end

ExperimentServer.start(4040)
```

In the preceding code snippet, we added a new option to the :gen_tcp.listen/2 call, reuseaddr. Using reuseaddr allows us to create a new listening socket even if the previously open socket isn't properly closed.

We also added the `listen/1` function in which we make a call to `:gen_tcp.accept/1` to create a connection socket. We further use that connection socket to send a `"Hello World"` response to the client before closing it.

Now, let's try to make a request to the server.

If you are using cURL to test the API, you might get the following response:

```
$ curl http://localhost:4040/
curl: (1) Received HTTP/0.9 when not allowed
```

So, I recommend using Wget for this specific request. Using Wget yields the expected results, as shown here:

```
$ wget -qO- http://localhost:4040/
Hello World
```

As you can see from this, we have successfully sent a `"Hello World"` response from the HTTP server. However, the response is using HTTP version 0.9, which isn't supported by many clients anymore.

> **cURL and HTTP/0.9**
>
> By default, using `send/2` sends an HTTP response using `HTTP/0.9`.
>
> However, cURL doesn't support HTTP versions less than 1.0. Therefore, in order to test the previous API call, we used Wget. The best way forward here isn't using Wget but updating our server to send an `HTTP 1.X` response.
>
> Let us update the `ExperimentServer` module to send an HTTP response with version 1.1, along with some headers and statuses. We will follow the HTTP response format we learned about in the first chapter.

So, let's update `ExperimentServer` to send a response using HTTP version 1.1:

---

experiment_server.exs

```elixir
defmodule ExperimentServer do
  # ..
  defp respond(connection_sock) do
    # Send a proper HTTP/1.1 response with status
    response = http_1_1_response("Hello World", 200)
    :gen_tcp.send(connection_sock, response)
    Logger.info("Sent response")

    :gen_tcp.close(connection_sock)
  end
```

```
  # Converts a body to HTTP/1.1 response string
  defp http_1_1_response(body, status) do
    """
    HTTP/1.1 #{status}\r
    Content-Type: text/html\r
    Content-Length: #{byte_size(body)}\r
    \r
    #{body}
    """
  end
end

ExperimentServer.start(4040)
```

Now that we are using HTTP version 1.1, we will be able to use `curl` to test our HTTP server:

```
$ curl http://localhost:4040/
Hello World%
```

We have a web server that only responds with `Hello World`. Now, let's update our web server to respond with `Hello World` to only a specific route, and with a `404` message to all other routes:

experiment_server.exs

```
defmodule ExperimentServer do
  # ..

  defp listen(listen_socket) do
    {:ok, connection_sock} = :gen_tcp.accept(listen_socket)
    {:ok, req} = :gen_tcp.recv(connection_sock, 0)

    {_http_req, method, {_type, path}, _v} = req

    Logger.info("Got request: #{inspect req}")
    respond(connection_sock, {method, path})

    listen(listen_socket)
  end

  defp respond(connection_sock, route) do
    response = get_response(route)

    :gen_tcp.send(connection_sock, response)
    Logger.info("Sent response")
```

```elixir
    :gen_tcp.close(connection_sock)
  end

  defp get_response(route) do
    {body, status} = body_and_status_for(route)
    http_1_1_response(body, status)
  end

  defp body_and_status_for({:GET, "/hello"}) do
    {"Hello World", 200}
  end

  defp body_and_status_for(_), do: {"Not Found", 404}

  defp http_1_1_response(body, status) do
    """
    HTTP/1.1 #{status}\r
    Content-Type: text/html\r
    Content-Length: #{byte_size(body)}\r
    \r
    #{body}
    """
  end
end

ExperimentServer.start(4040)
```

This will respond with 404 to any other path but /hello, for which it will respond with Hello World.

Let's try making a request to the /hello path. We should expect to get Hello World, but instead, we get a 404 response. This is because *we're using exact matches*, and our routes need to match the incoming request's path exactly. There are better ways of matching a request's path, which we will cover.

Now that we have a working web server module written using :gen_tcp, it's time to use that code to write our own web server package, which can be used with our web framework.

Writing an HTTP server package

Let's create a new goldcrest folder for all Goldcrest-related code, examples, and packages. Inside goldcrest, we can create our web server package by running the following command:

```
$ mix new goldcrest_http_server
```

Let's now define the `Goldcrest.HTTPServer` module, which will be our main web server module:

lib/goldcrest_http_server.ex

```elixir
defmodule Goldcrest.HTTPServer do
  @moduledoc """
  Starts a HTTP server on the given port

  This server also logs all requests
  """

  require Logger

  @server_options [
    active: false,
    packet: :http_bin,
    reuseaddr: true
  ]

  def start(port) do
    case :gen_tcp.listen(port, @server_options) do
      {:ok, sock} ->
        Logger.info("Started a webserver on port #{port}")

        listen(sock)

      {:error, error} ->
        Logger.error("Cannot start server on port #{port}:
                      #{error}")
    end
  end

  def listen(sock) do
    {:ok, req} = :gen_tcp.accept(sock)

    {
      :ok,
      {_http_req, method, {_type, path}, _v}
    } = :gen_tcp.recv(req, 0)

    Logger.info("Received HTTP request #{method} at
                 #{path}")

    respond(req, method, path)
```

```
      listen(sock)
  end

  defp respond(req, method, path) do
    # This part is different for different applications

    :gen_tcp.send(req, resp_string)

    Logger.info("Response sent: \n#{resp_string}")

    :gen_tcp.close(req)
  end
end
```

Most of the code in the preceding snippet was taken from the ExperimentServer module written earlier in this chapter. We have purposefully left the respond/3 function incomplete because we know that apps will respond differently to different requests. In order to make that part of the server configurable, we can delegate getting a response to a responder module that can be configured at the application level.

We can do that by defining a responder/0 function that returns the configured responder module and makes a call to a resp/3 function in the configured module:

lib/goldcrest_http_server.ex

```
defmodule Goldcrest.HTTPServer do
  # ..

  def respond(req, method, path) do
    %Goldcrest.HTTPResponse{} = resp =
      responder().resp(req, method, path)
    resp_string = Goldcrest.HTTPResponse.to_string(resp)

    :gen_tcp.send(req, resp_string)

    Logger.info("Response sent: \n#{resp_string}")

    :gen_tcp.close(req)
  end

  defp responder do
    Application.get_env(:goldcrest_http_server, :responder)
  end
end
```

Adding these functions (`respond/3` and `responder/0`) allows us to delegate the response generation to a module outside of this server.

You might have also noticed that I am pattern matching on a new struct, `Goldcrest.HTTPResponse`. Let's define that next.

Defining an HTTP response struct

It's always a good idea to define a struct to represent the structure of data you are expecting to work with. In this package, we will be working with HTTP responses, as our responder modules will be responding with them, and we will also be translating them into stringified documents to respond with. So, let's go ahead and define the struct module:

lib/goldcrest_http_response.ex

```
defmodule Goldcrest.HTTPResponse do
  defstruct headers: %{}, body: "", status: 200

  @type t :: %__MODULE__{
    headers: map(),
    body: String.t(),
    status: integer()
  }

  @http_version 1.1

  def to_string(
    %__MODULE__{
      status: status, body: body
    } = http_response
  ) do
    """
    HTTP/#{@http_version} #{status}\r
    #{Enum.join(headers_string(http_response), "\r\n")}
    \r
    #{body}
    """
  end

  defp headers_string(http_response) do
    http_response
    |> headers
    |> Enum.map(fn {key, value} ->
      "#{key}: #{value}"
```

```elixir
    end)
  end

  defp headers(%__MODULE__{body: body, headers: headers}) do
    headers = prep_headers(headers)

    body
      |> default_headers()
    |> Map.merge(headers)
  end

  defp default_headers(body) do
    %{
      "content-type" => "text/html",
      "content-length" => byte_size(body)
    }
  end

  defp prep_headers(headers) do
    headers
    |> Enum.map(fn {key, value} ->
      {String.downcase(key), value}
    end)
    |> Enum.into(%{})
  end
end
```

The module defined in the preceding code snippet defines a struct with the following fields:

- headers: Defaults to an empty map type. We chose the map type here because HTTP headers are generally unique and need to have unique keys.

- body: Defaults to an empty string. Represents the body of a response.

- status: Defaults to 200 (success).

On top of defining the struct, we have also defined a helper function that converts an HTTPResponse struct to the string version of the document, which can be used to respond with. We are also lower-casing all our header keys because HTTP headers are case-insensitive, and lower-casing them before validating is a quick way to ensure that they are unique.

Now, we can use this module in our HTTP server to generate response documents.

Validating whether a responder is configured

Next, we want to validate whether an HTTPServer module is properly configured to use a responder since it is a requirement to use the HTTPServer module. We can do this by checking for the presence of a responder configuration at the time of starting the server. So, in the start/1 function, let's validate the presence of a responder before creating a listening socket:

lib/goldcrest_http_server.ex

```
defmodule Goldcrest.HTTPServer do
  # ..

  def start(port) do
    ensure_configured!()

    case :gen_tcp.listen(port, @server_options) do
      {:ok, sock} ->
        Logger.info("Started a webserver on port #{port}")

        listen(sock)

      {:error, error} ->
        Logger.error("Cannot start server on port #{port}:
                      #{error}")
    end
  end

  defp ensure_configured! do
    case responder() do
      nil -> raise "No `responder` configured for
              `goldcrest_http_server`"
      _responder -> :ok
    end
  end

  defp responder do
    Application.get_env(:goldcrest_http_server, :responder)
  end
  # ..
end
```

Now, if we try to start the server without giving it a responder in the application's configuration, it will raise a RuntimeError.

Creating responder behavior

Since we expect the applications using this HTTP server to define a `responder` module, it is a good idea to define a `behaviour` module for it.

Behaviors in Elixir

In Elixir, a behavior is a way of defining a set of functions that a module should implement to obey that behavior.

We use typespecs to define a behavior using the `@callback` module attribute.

On the implementation module, we can register a behavior using the `@behaviour` module attribute. This will allow the compiler to warn us if a function that was expected to be implemented wasn't implemented.

Behaviors are a great way to define a shared API for a set of modules by defining a set of function specifications and having the ability to check whether those functions were actually implemented.

Let's now define a `behaviour` module for a responder. We know that a responder must implement the `resp/3` function, which takes a request, an HTTP method, and a path as a string to return an `HTTPResponse` struct. We can also define a `method()` type for all the HTTP methods, to better define the function specifications:

lib/goldcrest_http_server/responder.ex

```
defmodule Goldcrest.HTTPServer.Responder do
  @type method :: :GET | :POST | :PUT | :PATCH | :DELETE

  @callback resp(term(), method(), string()) ::
    Goldcrest.HTTPResponse.t()
end
```

Another useful module here would be a set of helper functions that help us write the `resp/3` function. In this case, those helper functions could enable us to create and update an `HTTPResponse` struct. Doing this will keep the code in a `responder` implementation easier to read and maintain.

So, let's define a `ResponderHelper` module that can be imported into a `responder` implementation module:

lib/goldcrest_http_server/responder_helpers.ex

```
defmodule Goldcrest.HTTPServer.ResponderHelpers do
  def http_response(body) do
    %Goldcrest.HTTPResponse{body: body}
  end
```

```
def put_header(%{headers: headers} = resp, key, value) do
  headers = Map.merge(headers, %{String.downcase(key) =>
    value})
  %{resp | headers: headers}
end

def put_status(resp, status) do
  %{resp | status: status}
end
end
```

In the preceding code snippet, we define three functions:

- `http_response/1`: This function creates a new HTTP response struct with the given body (as a function argument).

- `put_header/3`: This function updates an HTTP response with a new set of headers, ensuring they're unique. This is done by downcasing the header key and calling `Map.merge/2` on the headers. This ensures there's only one value corresponding to a key.

- `put_status/2`: This function updates the status of the HTTP response struct with the given status.

The preceding three functions are meant to provide a simple interface to build and update an HTTP response struct, which is the expected result of a responder module. This takes away the need to manipulate the `HTTPResponse` struct inside a responder module, hence making its implementation internal to the server. In order to ensure our HTTP server is functional, we will be learning how to test it in the next section.

Testing the HTTP server

Now that we have all the ingredients to write an HTTP server, it's time to write some tests. I will only cover some necessary tests in this chapter. If you are curious about which other tests we could write, feel free to check out the code for `goldcrest_http_server`, linked in the book's GitHub repository.

In order to test our HTTP server, we will first need to make two changes to the `mix.exs` file, as follows:

- Add Finch to the list of dependencies: This allows us to make HTTP calls to our server's endpoints.

- Update `elixirc_paths`: This updates the list of paths that need to be compiled as part of the project. Moreover, by defining those paths based on the Mix environment, Mix allows us to compile support modules in the test environment, one of which would be a `TestResponder` module.

Let's go ahead and make these changes:

mix.exs

```elixir
defmoduleGoldcrestHttpServer.MixProject do
  use Mix.Project
  # ..

  def project do
    [
      # ..
      elixirc_paths: elixirc_paths(Mix.env()),
      deps: deps()
    ]
  end
  # ..

  defp elixirc_paths(:test), do: ["lib", "test/support"]
  defp elixirc_paths(_), do: ["lib"]
  # ..
end
```

Now that we have added the test/support folder to the elixirc path for the test environment, we can define the TestResponder module in test/support/test_responder.ex. We can use this responder to test API calls.

In this responder module, we will define a route for /hello, which responds with Hello World with a 200 status and has the ability to respond with 404 for any other requests:

test/support/test_responder.exs

```elixir
defmodule Goldcrest.TestResponder do
  import Goldcrest.HTTPServer.ResponderHelpers

  @behaviour Goldcrest.HTTPServer.Responder

  @impl true
  def resp(_req, method, path) do
    cond do
      method == :GET && path == "/hello" ->
        "Hello World"
        |> http_response()
        |> put_header("Content-type", "text/html")
        |> put_status(200)
```

```
        true ->
          "Not Found"
          |> http_response()
          |> put_header("Content-type", "text/html")
          |> put_status(404)
      end
    end
end
```

Note how in the preceding code snippet, `ResponderHelpers` make building and modifying the `HTTPResponse` struct more ergonomic.

Now, we can use `Finch` and `TestResponder` to test our HTTP server:

test/goldcrest_http_server_test.exs

```
defmodule Goldcres.HTTPServerTest do
  use ExUnit.Case, async: false

  setup_all do
    Finch.start_link(name: Goldcrest.Finch)

    :ok
  end

  describe "start/2" do
    setup tags do
      responder = tags[:responder]

      old_responder =
        Application.get_env(
          :goldcrest_http_server,
          :responder
        )

      Application.put_env(
        :goldcrest_http_server,
        :responder,
        responder
      )

      on_exit(fn ->
        Application.put_env(
          :goldcrest_http_server,
          :responder,
```

```elixir
        old_responder
      )
    end)


    :ok
end

@tag responder: nil

test "raises when responder not configured" do
  assert_raise(
    RuntimeError,
    "No `responder` configured for
     `goldcrest_http_server`",
    fn -> Goldcrest.HTTPServer.start(4041) end
  )
end

@tag responder: Goldcrest.TestResponder

test "starts a server when responder is configured" do
  Task.start_link(fn ->
    Goldcrest.HTTPServer.start(4041)
  end)

  {:ok, response} =
    :get
    |> Finch.build("http://localhost:4041/hello")
    |> Finch.request(Goldcrest.Finch)

  assert response.body == "Hello World"
  assert response.status == 200
  assert {"content-type", "text/html"} in
    response.headers

  {:ok, response} =
    :get
    |> Finch.build("http://localhost:4041/bad")
    |> Finch.request(Goldcrest.Finch)

  assert response.body == "Not Found"
  assert response.status == 404
  assert {"content-type", "text/html"} in
    response.headers
end
```

```
      end
   end
```

In the preceding module, we have tested the following two use cases:

- When no responder is configured: In this case, we expect a RuntimeError saying a responder configuration is required to start the server

- When TestResponder is configured: In this case, we expect the /hello path to respond with a 200 status and any other request to respond with 404

We have used a setup block that sets a responder configuration. We then use ExUnit tags to pass a responder corresponding to each test.

This test can be run using the following command:

```
$ mix test
```

Assuming both the test cases pass, it's time to test our HTTP server with a sample application.

Running the HTTP server with a sample application

In order to test the HTTP server package with a sample application, we can create a new application with the --sup option so that we can start the HTTP server with the application. This can be done by running the following command in the root of the HTTP server package itself:

```
$ mix new test_goldcrest_http_server --sup
```

Next, we can add goldcrest_http_server to mix dependencies using relative paths. This is a great way to test a package since you can be sure to always use the *current* version of the package while running your tests:

mix.exs

```
defmodule TestGoldcrestHTTPServer.MixProject do
   # ..

   defp deps do
      [
        {:goldcrest_http_server, path:
          "../goldcrest_http_server"}
      ]
   end
end
```

Next up is defining a `Responder`. We can define a `resp/3` function just as with the `TestResponder` module and define the same two routes:

lib/test_goldcrest_http_server/responder.exs

```
defmodule TestGoldcrestHTTPServer.Responder do
  import Goldcrest.HTTPServer.ResponderHelpers

  def resp(_req, method, path) do
    cond do
      method == :GET && path == "/hello" ->
        http_response("Hello World")
        |> put_header("Content-type", "text/html")
        |> put_status(200)
      true ->
        http_response("Not Found")
        |> put_header("Content-type", "text/html")
        |> put_status(404)
    end
  end
end
```

We will also need to configure the responder to be the `Responder` module we created in the last code snippet:

config.exs

```
import Config

config :goldcrest_http_server,
  responder: TestGoldcrestHTTPServer.Responder
```

In order to add the HTTP server to a supervision tree, it's best to define a `child_spec/1` function. This allows us to add `HTTPServer` as a long-running process to the application's supervision tree:

lib/goldcrest_http_server.ex

```
defmodule Goldcrest.HTTPServer do
  # ..

  def child_spec(init_args) do
    %{
      id: __MODULE__,
      start: {__MODULE__, :start, init_args}
    }
```

end
end

The `child_spec/1` function in the previous snippet defines a way to start the `HTTPServer` module when it's added to a supervision tree, by just calling the `Goldcrest.HTTPServer.start/1` function.

Now, we can add the HTTP server to the sample application's supervision tree with port `4001`:

lib/test_goldcrest_http_server/application.ex

```elixir
defmodule TestGoldcrestHTTPServer.Application do
  @moduledoc false

  use Application

  @port 4001

  @impl true
  def start(_type, _args) do
    children = [
      {Goldcrest.HTTPServer, [@port]}
    ]

    opts = [
      strategy: :one_for_one,
      name: TestGoldcrestHTTPServer.Supervisor
    ]

    Supervisor.start_link(children, opts)
  end
end
```

Now, if we start the application, it will also start the HTTP server with the routes defined in the responder. We can start the application using the following command:

```
$ mix run --no-halt
```

We can confirm that the HTTP server is running by hitting the `/hello` endpoint and confirming that we get a `Hello World` response:

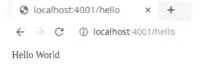

Figure 2.1 – Successful HTML response in a browser

We can also confirm to see that other routes produce a 404 response:

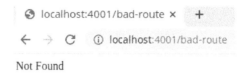

Figure 2.2 – Not Found HTML response in a browser

Now that our server is properly responding to HTTP requests with proper HTML documents, it's time to learn how and why to make it concurrent.

Adding concurrency to the server

In this section, we will learn how to add concurrency to the HTTP server. First, let us see why we need to add concurrency to our server. In order to understand that better, let's add another route to the Responder module, which performs a long-running operation. We can simulate that here by just sleeping for five seconds:

lib/test_goldcrest_http_server/responder.exs

```
defmodule TestGoldcrestHTTPServer.Responder do
  import Goldcrest.HTTPServer.ResponderHelpers

  def resp(_req, method, path) do
    cond do
      # ..

      method == :GET && path == "/long-operation" ->
        # sleep for 5 seconds
        :timer.sleep(5000)

        http_response("Hello World LONG LONG")
        |> put_header("Content-type", "text/html")
        |> put_status(200)

      true ->
        http_response("Not Found")
        |> put_header("Content-type", "text/html")
        |> put_status(404)
    end
  end
end
```

Upon restarting the application, if we try to first hit the /long-operation endpoint promptly followed by the /hello endpoint, we will see that even though /hello isn't a costly operation, it takes a while to respond.

This is because our server is processing all of the requests sequentially. This means that when it received the /hello request, it was still waiting to process the first /long-operation request. This isn't a very scalable design as a single long-running endpoint could potentially slow down the entire application:

Figure 2.3 – Failed long response in a browser

We can fix this problem by using Elixir's concurrency model. By spawning a new process every time the server processes a request, we can ensure that even if one request is still being processed, another process can pick up a new request. We can do this by spawning a new process when calling the respond/3 function. This would also require us to make the respond/3 function public so that it can be called with the spawn/3 command:

---

lib/goldcrest_http_server.ex

```elixir
defmodule Goldcrest.HTTPServer do
  # ..

  def listen(sock) do
    {:ok, req} = :gen_tcp.accept(sock)

    {
      :ok,
      {_http_req, method, {_type, path}, _v}
    } = :gen_tcp.recv(req, 0)

    Logger.info("Received HTTP request #{method} at
      #{path}")

    spawn(__MODULE__, :respond, [req, method, path])

    listen(sock)
  end
end
```

Now, if you try to follow the same process as earlier of quickly following up a request to the /long-operation endpoint with /hello, you will see that even though the first request hasn't finished responding, the second request is completed:

Figure 2.4 – Successful long response in a browser

The ease with which this can be achieved and managed on the BEAM virtual machine is one of the main reasons why the Erlang and Elixir ecosystems make great candidates for web server development. Now that we have tried our HTTP server with a web application, let's learn how to run it in detached mode similar to Phoenix.

Bonus – starting a detached HTTP server

Running a detached server is quite useful, especially in scenarios where you need to run a detached console for debugging purposes while the app is running. We can confirm this by running an iex console with the following command:

```
$ iex -S mix
[info]   Started a webserver on port 4001
```

Running the iex command should print the preceding info log. Even though you started an iex shell, you won't have access to the shell or any other process in the supervision tree. In order to fix this, we can use Task.start_link/3 and run the HTTP server process as a Task.

Let's update the child_spec/1 function in the HTTP server to start a Task:

lib/goldcrest_http_server.ex

```
defmodule Goldcrest.HTTPServer do
  # ..

  def child_spec(init_args) do
    %{
      id: __MODULE__,
      start: {
        Task,
        :start_link,
        [fn -> apply(__MODULE__, :start, init_args) end]
      }
```

```
      }
   end
   # ..
end
```

Now, running the `iex` shell again should start the server without stopping you from accessing the shell:

```
$ iex -S mix
Interactive Elixir (1.11.3) - press Ctrl+C to exit (type h() ENTER for
help)
[info]   Started a webserver on port 4001
iex(1)>
```

Summary

In this chapter, we covered the basics of the `:gen_tcp` module. We saw how `:gen_tcp` allows us to create a connection using a TCP/IP socket, and how we can use it to write our own HTTP server. We then wrapped the HTTP server into a package while making the implementation for its response configurable using an application environment.

Later, we learned how to test our HTTP server using the Finch HTTP client, and made the server concurrent by spinning up a new process for every HTTP request, similar to what Cowboy does. We also updated our HTTP server to have the ability to run in detached mode just like Phoenix and Cowboy.

Now that we have our own HTTP server up and running, we have to make it easier to define routes and responses. For a complex web application, matching the routes the way we did in `Responder` modules wouldn't be very helpful. Luckily, the Elixir community has a `Plug` package that helps us with the entire request-response data manipulation process; we will be covering this in the next chapter.

Exercise

To solidify what you've learned in this chapter, you can attempt to write tests for the `test_goldcrest_http_server` sample app.

Part 2:
Router, Controller, and View

In this part, you will learn how to build the core part of the web framework, Goldcrest, which includes the router, controller, and view.

This part includes the following chapters:

- *Chapter 3, Defining Web Application Specifications Using Plug*
- *Chapter 4, Working with Controllers*
- *Chapter 5, Adding Controller Plugs and Action Fallback*
- *Chapter 6, Working with HTML and Embedded Elixir*
- *Chapter 7, Working with Views*

3
Defining Web Application Specifications Using Plug

In the previous chapters, we took a deep dive into Cowboy to learn about some of its fundamentals. Then, using that knowledge, we built a web server using `:gen_tcp`. However, in both those chapters, we didn't deal with any sort of complex route matching or other request formats. We will now learn how to accomplish that using the **Plug** package. In this chapter, we will learn how Plug works, understand some of its main components, and how we can use them to make our web server more usable.

By the end of this chapter, you will know the ins and outs of the Plug package and how Plug enables us to use Cowboy efficiently. You will also learn how to make Plug work with the HTTP server we built in the previous chapter.

In this chapter, we will cover the following topics:

- Philosophy of Plug
- Components of Plug
- The `Plug.Conn` struct
- What happens when you send a response using Plug?
- Function plugs and module plugs
- Using Plug to write a router for our requests
- Using the `Plug` adapter for Cowboy
- Creating a `Plug` adapter for an HTTP server
- Using Plug along with the HTTP server we built in *Chapter 2*
- Simplifying the interface for calling our new `Plug` adapter

Technical requirements

This chapter contains a lot of code snippets that are best understood by following along. Just like in the previous chapters, we will use Elixir 1.12.x and Erlang 23.2.x. We will also be using cURL to test our HTTP server and running multiple commands on the terminal to both run our server and test it using cURL. So, a way to multiplex your terminal would be useful.

The code examples for this chapter can be found at `https://github.com/PacktPublishing/Build-Your-Own-Web-Framework-in-Elixir/tree/main/chapter_03`

What is Plug?

Plug is an Elixir module that allows us to easily define the specifications of a web application. If you are from the Ruby world, you can think of Plug as somewhat like Rack, which allows us to define middleware to read and transform requests/responses in a web application.

Plug allows us to add middleware-like behavior by defining a set of composable operations on a connection struct, `Plug.Conn`. This allows us to define behavior in the form of a plug, which can transform the attributes of a `Conn`. Unlike in Rack, `Conn` contains information related to both the request and the response. This, along with Elixir's `|>` operator, allows us to define the entire request-response cycle in a single pipeline of composable operations:

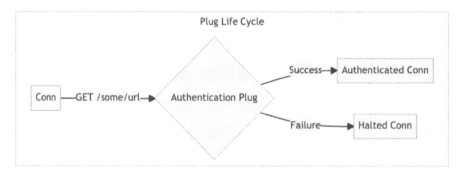

Figure 3.1 – Plug life cycle

Representing the state of the connection with the Plug. Conn struct

Plug uses the `Plug.Conn` struct to represent the state of a connection at a point in time. This includes information related to the request, cookies, response, and status codes.

This struct fits well into the philosophy of using plugs, which involves using a plug in a functional way to transform the attributes of a `Plug.Conn` struct. This also means that a plug receives a `Plug.Conn` with a state denoting that the response has already been set or that the connection was halted; this plug ignores that connection.

Some of the other useful information that `Plug.Conn` tracks are as follows:

- `req_headers`: Represents the downcased version of request headers. This is super helpful for operations such as authentication or caching.
- `params`: Denotes merged map parameters sent as path params and body params.
- `cookies`: Request and response cookies.
- `status`: Response status.
- `assigns`: A place to store user data and share it across plugs.
- `private`: Another way to share data across plugs. This is usually reserved for libraries and frameworks that leverage plugs, such as Phoenix.
- `adapter`: This indicates how `Plug.Conn` will be used. This field is generally used by HTTP servers to handle events such as sending a response or sending a file. Later in this chapter, we will define an adapter for the HTTP server we created in *Chapter 2*.

The `Plug.Conn` module also provides us with helper functions to update the state of the `Plug.Conn` struct:

- `assign/3`: Used to store a value corresponding to a key in a connection to update the `assigns` map. The `assigns` map is used to pass information between plugs in a pipeline of plugs.
- `halt/1`: Sets the `halted` state for a `Plug.Conn` struct to `true`. This prevents any further plugs from being called in the pipeline.
- `send_resp/3`: Sends a response with the given HTTP status and the body.

What happens when you send_resp/3?

When we call `Plug.Conn.send_resp/3`, Plug first checks the state of the `Plug.Conn` struct. If the state is not set (`nil`), it raises an error, indicating that the state is `nil`, while if the state is set to `:sent` (or anything else), it raises `AlreadySentError`. If the state is `:set`, indicating that the connection is ready to be sent, it calls the `send_resp/4` function of the adapter module set in the `Plug.Conn` struct. In the case of the `Plug.Cowboy` adapter, the response gets sent using the Cowboy web server. This function also sends a message to the process that owns the connection, indicating that a response was sent. Finally, the function returns a connection with the `:sent` state.

A very similar workflow occurs for other functions, including `put_status/2`, `send_file/4`, `send_chunked/2`, and others.

Now that we know more about `Plug.Conn`, let's learn how to define a plug that manipulates the `Plug.Conn` struct.

Defining a plug

A plug takes two forms: a **function plug** and a **module plug**.

A function plug simply takes a connection with a set of options as a keyword list and returns a new connection. For example, the following `Plug` function just takes a connection, checks if it is authorized, and returns a halted connection if it is not:

```
import Plug.Conn

def authorization_plug(conn, opts) do
  if is_authorized?(conn) do
    conn
  else
    conn
    |> put_resp_content_type("text/plain")
    |> put_status(401)
    |> halt()
  end
end
```

Function plugs like this one are great for one-time use with not many use cases. However, if you want a plug to be called on multiple pipelines or to be part of a supervision tree, it's better to use a module plug.

A `Plug` module can be used to transform a connection struct based on the given options and parameters. For a module to behave like a plug, it needs to implement two functions:

- `init/1`: This function is called to initialize the options for a `Plug` module before invoking it using the `call/2` function. This function is generally used for validating options or transforming options into what the `call/2` function expects. The options returned by `init/2` are passed to `call/2` as the second argument.

- `call/2`: This function is called every time a `Plug` module is invoked. The first argument to `call/2` is a `Plug.Conn` struct, while the second argument is a list of options (usually returned by the `init/1` callback). This function should return a new `Plug.Conn` struct with the appropriate state indicating that this `Plug` module was indeed called.

Now, let's define a new `Plug` module with these two functions:

example_plug.exs

```
Mix.install([
  {:plug_cowboy, "~> 2.0"}
])

defmodule GreeterPlug do
  @moduledoc """
  This plug greets the world based on the given greeting
  """

  import Plug.Conn

  @doc """
  Validates options.

  Ensures a greeting is present
  """
  def init(options) do
    greeting = Keyword.get(options, :greeting, "Hello")
    [greeting: greeting]
  end

  @doc """
  Returns a conn with a greeting response
  """
  def call(conn, greeting: greeting) do
    conn
    |> put_resp_content_type("text/plain")
    |> send_resp(200, "#{greeting} World!")
  end
end
```

The preceding `Plug` will be responsible for adding a response based on the given options.

In the `GreeterPlug` module, we expect to get a greeting as an option, which otherwise defaults to `"Hello"`. This is handled by the `init/1` function, which translates the options into a keyword list with only one key, `:greeting`.

Since `init/1` always returns a keyword with the `:greeting` key, the `call/2` function matches on the `greeting: greeting` pattern as the second argument. It then returns a conn with a response stating `"#{greeting} World!"`.

To use this plug, we must use it with Cowboy's `Plug` implementation. To do that, we can add a `:plug_cowboy` dependency to our script by leveraging the `Mix.install/2` function:

example_plug.exs

```
defmodule GreeterPlug do
  # ..
end

%{start: {mod, fun, args}} =
  Plug.Cowboy.child_spec(
    scheme: :http,
    plug: GreeterPlug,
    options: [port: 4040]
  )

apply(mod, fun, args)
IO.puts "Listening on port 4040"

# Run this server for 10 seconds
:timer.sleep(10_000)
```

In the preceding file, we use the `Plug.Cowboy.child_spec/1` function to get the **Module, Function, and Arguments (MFA)** of the function we can call to start the HTTP server. We pass `GreeterPlug` as our `plug` and `port: 4040` as `options` to Cowboy to get the correct arguments. We have also added a `:timer.sleep/1` function to give us 10 seconds to test out our HTTP server.

Running the preceding file will start an HTTP server for 10 seconds on port `4040`:

```
$ elixir example_plug.exs
Listening on port 4040
```

To test the preceding plug, we can make a request to `localhost:4000`. In a separate terminal, we can use `wget` to make an HTTP request:

```
$ wget -qO- http://localhost:4000
Hello World
```

We can now see that the `GreeterPlug` module adds a text response of `Hello World` to the conn struct, which gets added to the final response from the HTTP server. So, that's how you define a `Plug` module. On top of providing interfaces to define new `Plug` modules, the Plug package itself comes with a few pre-defined modules that make life easier while working with HTTP requests and responses. One such module is `Plug.Router`, which provides a tidy interface for matching HTTP requests based on the parameters of the request.

Understanding Plug.Router

Now that we know how a `Plug` module works, let's leverage one of the predefined plugs, `Plug.Router`, to add functionality to our HTTP server. At the end of *Chapter 2*, we saw that to use our HTTP server for a proper REST API, we need a way to route incoming requests based on different paths and HTTP verbs. Cowboy has an internal way of routing the requests, but `Plug.Router` allows us to do that in a very simple DSL. Another advantage of using `Plug.Router` to route our requests, instead of using Cowboy directly, is that we can start using plugs earlier in our request-response cycle, even before they get to Cowboy. Here, we can only rely on Cowboy to send a response.

So, let's start by defining a new plug, `ExampleRouter`:

example_router.exs

```
Mix.install([
  {:plug_cowboy, "~> 2.0"}
])

defmodule ExampleRouter do
  use Plug.Router

  plug :match
  plug :dispatch

  get "/greet" do
    send_resp(conn, 200, "Hello World")
  end

  match _ do
    send_resp(conn, 404, "Not Found")
  end
end

%{start: {mod, fun, args}} =
  Plug.Cowboy.child_spec(
    scheme: :http,
    plug: ExampleRouter,
    options: [port: 4040]
  )

apply(mod, fun, args)
IO.puts "Listening on port 4040"
```

```
# Run this server for 10 seconds
:timer.sleep(10_000)
```

The preceding file is very similar to ./example_plug.exs, where we use a Plug module, along with the Plug.Cowboy adapter, to start an HTTP server at a given port. However, with Router, we can not only send an HTTP response but also match the path and the HTTP verb of the incoming request.

Plug.Router defines a pipeline of plugs and expects two plugs to be in the pipeline:

- :match: This plug is responsible for matching an incoming request's path and verb to a route defined later in the module. When a route is matched, it sends the connection to the :dispatch plug.

- :dispatch: This plug is responsible for calling an implementation for a matched request.

In the preceding module, we can see that all the requests to /greet will get a response of 200 with a body of "Hello World". Any other requests will be responded to with a 404.

We can also see how simple the interface for Plug.Cowboy is. We can simply call the child_spec/1 function, which returns MFA details that can be called using Kernel.apply/3 to start a server linked with the running process. And since the HTTP server is linked to the running process, we have added a sleep for 10 seconds to give us enough time to test out the HTTP server.

Now that we have learned how to use Plug with Cowboy, it's time to define an interface so that we can use Plug with our HTTP server.

Plugifying our HTTP server

To leverage the power of Plug for routing, error handling, and other uses, let's define a Plug adapter for the Goldcrest.HTTPServer package.

Let's start by defining a new package, plug_goldcrest_http_server:

```
$ mix new plug_goldcrest_http_server --module Plug.Goldcrest.
HTTPServer
```

Now, let's add :goldcrest_http_server and plug to our Mix dependencies:

lib/plug/goldcrest/http_server/conn.ex

```
defmodule Plug.Goldcrest.HTTPServer.MixProject do
  use Mix.Project
  # ..
  defp deps do
    [
```

```
      {:goldcrest_http_server, "~> 0.0.1"},
      {:plug, "~> 1.12.1"}
    ]
  end
end
```

Now, we can fetch the dependencies by running the following command:

```
$ mix deps.get
```

In this newly generated mix project, we can start by defining an adapter for `Plug.Conn`. Let's call the module `Plug.Goldcrest.HTTPServer.Conn`:

To define an adapter, our `Plug.Goldcrest.HTTPServer.Conn` module needs to implement a few callbacks defined in `Plug.Conn.Adapter`. Some of those callbacks are as follows:

- `send_resp/4`: Used to send a given response, along with its status and headers
- `send_file/6`: Used to send a given file, along with its status and headers
- `send_chunked/3`: Used to send chunked responses to the client, along with their statuses and headers

To keep this concise, we will only be implementing the `send_resp/4` function. This function will be further delegated to our HTTP server helpers and will utilize `:gen_tcp` to send the final response, along with the given status and the headers.

Now, let's write our adapter:

lib/plug/goldcrest/http_server/conn.ex

```
defmodule Plug.Goldcrest.HTTPServer.Conn do
  import Goldcrest.HTTPServer.ResponderHelpers

  @moduledoc false
  @behaviour Plug.Conn.Adapter

  @impl true
  def send_resp({req, method, path}, status, headers, body)
  do
    resp_string =
      body
      |> http_response()
      |> apply_headers(headers)
      |> put_status(status)
      |> Goldcrest.HTTPResponse.to_string()
```

```
      :gen_tcp.send(req, resp_string)

      :gen_tcp.close(req)

      {:ok, nil, {req, method, path}}
    end

    defp apply_headers(resp, headers) do
      Enum.reduce(headers, resp, fn {k, v}, resp ->
        put_header(resp, k, v)
      end)
    end
  end
```

In the preceding code, we imported the helper module we defined in *Chapter 2*, `Goldcrest`. `HTTPServer.ResponderHelpers`. This module contains the `http_response/1`, `put_header/3`, and `put_status/2` functions. We used these imported functions in `send_resp/3` to construct our HTTP response with proper headers and statuses.

Once we had the final response string, we used the request socket, which was passed along with `method` and the request's `path` to `send_resp/3`, to send the final response to the client using `:gen_tcp.send/2`. We followed that up by closing the connection socket using `:gen_tcp.close/1` and responded with `{:ok, nil, {req, method, path}}`. In this way, `req`, `method`, and `path` can be used by other upcoming plugs.

Keep in mind that the state of the connection is set to `:sent`, which will prevent other plugs from being called after a response is sent via `send_resp/3`. All of this is handled by the `Plug` package itself.

Now, we need a way to initialize the plug operations for our HTTP server. To do that, we first need a way to build a `Plug.Conn` struct from a request received using `Goldcrest.HTTPServer`.

We know that the `adapter` property for that `Plug.Conn` struct will contain `Plug.Goldcrest.HTTPServer`, so it can use functions such as `send_resp/3`, which we define in that module. And, since the first argument for `send_resp/3` is the second element of the tuple corresponding to the value of the `:adapter` field in the `Plug.Conn` struct, we will need to use `{Plug.Goldcrest.HTTPServer, {req, method, path}}` as the value for the `adapter` property for the `Plug.Conn` struct.

We can also use the `method` and `path` properties we get from `HTTPServer` to fill the `method` and `path` fields for the struct. We can use `:inet.sockname/1` to get the value for the `remote_ip` field, and we can split `path` using `Path.relative_to/2` and `Path.split/1` to get `path_info`, which needs to be a list of path components.

We will be skipping the `host` and `port` fields since we will not be using them in this chapter.

We will also be using default values for `scheme` as `:http`.

Finally, we will be using URI.parse/1 to get path and query information from the given request's URI.

So, let's define the conn_from_req function in the Plug.Goldcrest.HTTPServer module:

lib/plug/goldcrest/http_server.ex

```elixir
defmodule Plug.Goldcrest.HTTPServer do
  @moduledoc """
  Documentation for `Plug.Goldcrest.HTTPServer`.
  """

  def conn_from_req(req, method, path) do
    {:ok, {remote_ip, _}} = :inet.sockname(req)
    %URI{path: path, query: qs} = URI.parse(path)

    %Plug.Conn{
      adapter: {@adapter, {req, method, path}},
      host: nil,
      method: Atom.to_string(method),
      owner: self(),
      path_info: path |> Path.relative_to("/") |>
      Path.split(),
      port: nil,
      remote_ip: remote_ip,
      query_string: qs,
      req_headers: [],
      request_path: path,
      scheme: :http
    }
  end
end
```

The preceding function takes a request socket, req, an HTTP method, method, and a request path, path (containing the query string). It uses the :inet, URI, and Path module's functions to get information from the req socket and path to finally return a Plug.Conn struct that's ready to be used by our HTTP server.

We have also made this function public so that it can be used by other parts of the system in the future for testing, debugging, and writing plugs easily.

Now that we have a way to initialize a `Plug.Conn` struct from a request, we are ready to define an `init/4` function that can be used to initialize our plug:

lib/plug/goldcrest/http_server.ex

```
defmodule Plug.Goldcrest.HTTPServer do
  # ..
  def init(req, method, path, plug: plug, options: opts) do
    conn = conn_from_req(req, method, path)

    %{adapter: {@adapter, _}} =
      conn
      |> plug.call(opts)
      |> maybe_send(plug)

    {:ok, req, {plug, opts}}
  end

  defp maybe_send(%Plug.Conn{state: :unset}, _plug) do
    raise(Plug.Conn.NotSentError)
  end

  defp maybe_send(%Plug.Conn{state: :set} = conn, _plug) do
    Plug.Conn.send_resp(conn)
  end

  defp maybe_send(%Plug.Conn{} = conn, _plug), do: conn

  defp maybe_send(other, plug) do
    raise "Goldcrest adapter expected #{inspect(plug)} to "
    <>
          "return Plug.Conn but got: #{inspect(other)}"
  end
end
```

In the preceding code, we added the `init/4` function, which takes a request socket, a method, a path, and a plug as an option. The plug option could be a router, which can be used to route requests to a handler. The `init/4` function then calls the `conn_from_req/3` function to initialize a `Plug.Conn` struct from a request and starts the Plug pipeline with the given options. After that, `maybe_send` is called to make sure that any connections with state `:set` are sent using the `send_resp/3` callback defined earlier in this chapter. Finally, the function returns the connection socket, along with the plug and its options.

Now that we have defined adapters for our HTTP server, let's test them out using a plug. To do that, we will start by defining a new router:

temp/example_router.ex

```
defmodule ExampleRouter do
  use Plug.Router

  plug :match
  plug :dispatch

  get "/greet" do
    send_resp(conn, 200, "Hello World")
  end

  match _ do
    send_resp(conn, 404, "Not Found")
  end
end
```

The preceding router simply responds with "Hello World" when it receives a GET request at the /greet path and returns a 404 for everything else.

Now, let's open an iex shell with the mix project loaded:

```
$ iex -S mix
iex> Code.require_file("./temp/example_router.ex")
iex> options = {Plug.Goldcrest.HTTPServer, [plug: ExampleRouter,
options: []]}
iex(1)> Application.put_env(:goldcrest_http_server, :dispatcher,
options)
iex(2)> Goldcrest.HTTPServer.start(4040)

21:40:21.771 [info]   Started a webserver on port 4040

21:40:36.402 [info]   Received HTTP request GET at /greet
```

The preceding snippet loads the file that defines the ExampleRouter module. Then, it updates the :dispatcher environment for :goldcrest_http_server to the new plug module that we created earlier in this chapter. This allows our HTTP server to delegate the handling of the request to the Plug.Goldcrest.HTTPServer module. It then starts the Goldcrest HTTP server at port 4040.

While this server is running, in a separate shell, let's make an HTTP request to our server to test whether our router is working:

```
$ curl -v  http://localhost:4040/greet
*    Trying 127.0.0.1:4040...
* Connected to localhost (127.0.0.1) port 4040 (#0)
> GET /greet HTTP/1.1
> Host: localhost:4040
> User-Agent: curl/7.79.1
> Accept: */*
>
* Mark bundle as not supporting multiuse
< HTTP/1.1 200
< cache-control: max-age=0, private, must-revalidate
< content-length: 11
< content-type: text/html
<
* Excess found in a read: excess = 1, size = 11, maxdownload = 11,
bytecount = 0
* Closing connection 0
Hello World
```

The preceding request responds with "Hello World" and a 200 as expected. Now, let's make a request to a bad path to test our 404 matching logic:

```
$ curl -v  http://localhost:4040/other
*    Trying 127.0.0.1:4040...
* Connected to localhost (127.0.0.1) port 4040 (#0)
> GET /other HTTP/1.1
> Host: localhost:4040
> User-Agent: curl/7.79.1
> Accept: */*
>
* Mark bundle as not supporting multiuse
< HTTP/1.1 404
< cache-control: max-age=0, private, must-revalidate
< content-length: 9
< content-type: text/html
<
* Excess found in a read: excess = 1, size = 9, maxdownload = 9,
bytecount = 0
* Closing connection 0
Not Found
```

As expected, the request returns a 404 with "Not Found".

Next, we will make our plug easier to use by defining the child_spec/1 function in a similar way to how we defined Plug.Cowboy.child_spec/1, which can make it easier for us to start the HTTP server that's linked to the running process, similar to Plug.Cowboy. We can also define a function, start_linked_server/1, to start an HTTP server process linked to the current process. So, let's define the child_spec/1 and start_linked_server/1 functions:

lib/plug/goldcrest/http_server.ex

```
defmodule Plug.Goldcrest.HTTPServer do
  # ...
  def child_spec(plug: plug, port: port, options: options)
  do
    Application.put_env(
      :goldcrest_http_server,
      :dispatcher,
      {__MODULE__, [plug: plug, options: options]}
    )

    %{start: {__MODULE__, :start_linked_server, [port]}}
  end

  def start_linked_server(port) do
    Task.start_link(fn ->
      Goldcrest.HTTPServer.start(port)
    end)
  end
end
```

In the preceding code snippet, we added the child_spec/1 and start_linked_server/1 functions to the Plug.Goldcrest.HTTPServer module. The child_spec/1 function updates the :dispatcher environment with the given attributes (plug and options). Then, it returns a specification with the value for start as an MFA. The function referenced by the MFA is the new Plug.Goldcrest.HTTPServer.start_linked_server/1 function, with the port as the argument, which starts a Goldcrest.HTTPServer process as a child process linked to the current process.

Now, we can simply start our server by calling apply on the MFA returned by the child_spec/1 function. Let's test it out in iex:

```
$ iex -S mix
iex> Code.require_file("./temp/example_router.ex")
iex> opts = [plug: ExampleRouter, port: 4040, options: []]
iex> %{start: {mod, fun, args}} = Plug.Goldcrest.HTTPServer.child_
spec(opts)
iex> apply(mod, fun, args)
```

```
{:ok, #PID<0.232.0>}

22:42:49.877 [info]  Started a webserver on port 4040
```

Now that our server's plug is easy to use, we can use it in a similar way to `Plug.Cowboy`. With that, our HTTP server is ready to be used with a web framework.

Summary

In this chapter, we learned how to leverage the Plug library to define a router to route incoming requests to a specific handler. We also learned about the Plug philosophy and why using Plug for web applications makes sense. We then used `Plug.Conn.Adapter` behavior to define an adapter for the HTTP server we defined in the previous chapter. Finally, we wrapped up by defining a cleaner interface for using the plug.

In this chapter, we didn't define all the implementations for callbacks defined in the `Plug.Conn. Adapter` module. This was done to save time. We also didn't write automated tests for this chapter because most of what we wrote can be tested similar to how HTTP servers were defined in previous chapters. We do intend on writing a better testing interface for our HTTP server, but we will do that in the DSL design part of this book.

In the next chapter, we will use our knowledge of Plug to build a controller interface to manage the flow of an incoming HTTP request.

4

Working with Controllers

In the previous chapter, we learned how to use the Plug package to simplify a pipeline of operations that are applied to an HTTP connection, represented by the %Plug.Conn{} struct. We also learned how we can leverage the Plug.Router plug to handle a request matching an HTTP method and path. In this chapter, we will take our Plug knowledge to the next level by applying the same philosophy (of updating a connection struct in a pipeline) to implement a **controller**. We will also take a quick look at how Phoenix implements its controllers and take inspiration from that to build our own controllers. Before doing any of that, however, we will learn what controllers are in a **Model-View-Controller** (**MVC**) framework and understand some of their basics.

By the end of this chapter, you will have learned the ins and outs of controllers in Phoenix, and how Phoenix uses Plug and Cowboy to route HTTP requests to proper handlers. You'll also have learned the fundamentals of redirects and testing a request-response cycle in a web application.

In this chapter, we will cover the following key topics:

- What are controllers?
- Controllers in Phoenix
- Building our own controller
- Controllers as plugs
- Using our controller with plug_cowboy
- Using our controller with plug_goldcrest_http_server
- How do URL redirections work?
- Using Plug.Test
- Testing our controller

Technical requirements

This chapter has a lot of code snippets that are best understood by following along. Similar to earlier chapters, we will use Elixir 1.12.x and Erlang 23.2.x.

In this chapter, I will assume you've read through the previous chapters and followed along by writing your code snippets. It is particularly important that you understand how `plug` and `goldcrest_http_server` work (covered in *Chapter 2* and *Chapter 3*) to get the most out of this chapter.

Since we will explore building our own controller, any experience with writing controllers in a web framework such as Phoenix or Rails, although not needed, would still be helpful.

The code examples for this chapter can be found at `https://github.com/PacktPublishing/Build-Your-Own-Web-Framework-in-Elixir/tree/main/chapter_04`

What are controllers?

A controller is the component of a web application that accepts input and **controls** the flow of requests. This generally includes accepting a specific content type, authenticating, matching request parameters, and so on.

In an MVC framework, a controller is generally used as an entry point to the web application. The controller first matches a request based on its parameters (such as request method, route, and so on), and uses the model layer to fetch any requested data. Finally, it uses the views to present the data in a consumable way. In many modern MVC frameworks such as Phoenix, there also exists an additional component, the router, which takes on the responsibility of matching an incoming request to a controller for further handling.

The request-response cycle described is better shown in the following diagram:

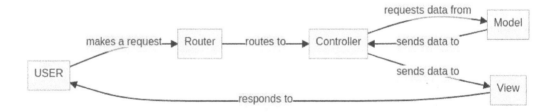

Figure 4.1 – The MVC flow

Now that you've looked at a brief overview of controllers in general, let's see how Phoenix implements them.

Controllers in Phoenix

In Phoenix, controllers are plugs. This is an extension of the plug philosophy that we covered in *Chapter 3*, where a web request goes through a pipeline of plugs that transform the connection, the `%Plug.Conn{}` struct, to indicate that a plug was called. Therefore, a router and a controller are both plugs responsible for transforming the connection and responding to the end user.

A controller is a module that contains several request handlers as functions. A Phoenix controller is called by a router based on the route of the incoming request. The request is also properly delegated to a particular request handler function in the controller depending on the route defined in the router. The controller functions that are responsible for handling a request are called `actions`. By default, a Phoenix controller's action has an arity of 2. It takes a `Plug.Conn` struct as the first argument and request parameters as a `Map` as the second argument.

Here's an example of a Phoenix controller:

```
defmodule ExampleWeb.PageController do
  use ExampleWeb, :controller

  def index(conn, params) do
    case Example.Page.load_data(params) do
      {:ok, data} ->
        render(conn, "index.html", data: data)

      {:error, error} ->
        conn
        |> put_flash(:error, error)
        |> redirect(to: "/")
    end
  end
end
```

In the preceding code, we can see that a controller is defined as a module that calls `use ExampleWeb, :controller`. We will talk more about the `use` macro in *Chapter 8*, but for the purposes of this chapter, all this does is enable this module to act like a controller by importing several functions and defining compile-time callbacks that would assist in request handling.

In the `index/2` function, we can see two arguments being passed. The first argument, `conn`, is a `Plug.Conn` struct, which holds a lot of information about the request, as we learned in the previous chapter. The second argument, `params`, is a `Map` that stores the request params (and more), allowing the application to fetch relevant information using `Example.Page.load_data/1`.

We can also see that there are two possible paths for the request to go based on the outcome of the `Example.Page.load_data/1` call. If the call is a success, it renders `"index.html"` with the loaded data; otherwise, it redirects to the root (`""`) with an error message. Both of these actions are performed by the helper functions, `render/3` and `redirect/2`, imported as a result of the use call on *line 2*.

Now that we know what a Phoenix controller module looks like, we can apply the same philosophy and structure to create a controller construct for our web framework.

Building a controller interface

In this section, we will learn what it takes to build a controller by leveraging the Plug package and everything we've learned about it.

To build our own controller, let's start by creating a new Mix project by running the following command:

```
$ mix new goldcrest
```

Now, our goal is to be able to write a controller that looks something like a Phoenix controller:

```
defmodule ExampleController do
  # .. TODO: use/import modules

  def greet(conn, _params) do
    conn
    |> put_status(200)
    |> render(:json, %{data: "hello world"})
  end
end
```

The previous controller defines one action, `greet/2`, which responds with a status of `200` along with the `{"data": "hello world"}` JSON body.

We have hidden the complexity of the `use` or `import` modules to introduce controller-like behavior to our module, but we will figure that part out in this section of the chapter.

To implement the preceding controller, we will first need to add `plug` as a dependency to the `goldcrest` app. We will also need to add `jason` as a dependency to decode and encode the JSON body for the request-response cycle:

mix.exs

```
defmodule Goldcrest.MixProject do
  # ...
  defp deps do
    [
```

```
      {:jason, "~> 1.2.2"},
      {:plug, "~> 1.12.1"}
    ]
  end
end
```

We can then run $ mix deps.get to fetch those dependencies for our Mix project.

To test our controller, we will need a router and an HTTP server.

Let's start by creating a router using the Plug.Router module:

lib/goldcrest/support/example_router.ex

```
defmodule Goldcrest.ExampleRouter do
  use Plug.Router

  plug :match
  plug :dispatch

  match _ do
    send_resp(conn, 404, "Not Found")
  end
end
```

The preceding router matches all the incoming requests and sends a 404 response indicating that no route has been implemented.

Now, let's create a new route, /greet, which we can direct to a controller plug:

lib/goldcrest/support/example_router.ex

```
defmodule Goldcrest.ExampleRouter do
  use Plug.Router
  # ..
  get "/greet" do
    Goldcrest.ExampleController.call(conn, action: :greet)
  end
  # ..
end
```

As you can see in the preceding code, we have added a new route, GET /greet, which dispatches any matching requests to the Goldcrest..ExampleController plug module. For now, we're explicitly invoking the controller like a plug using the call/2 function with :action as the option with the :greet value, indicating that this request should go to a Goldcrest.ExampleController. greet/2 controller action.

Now that we know what our controller should look like, let's go ahead and write it. We will need to define two functions in our controller:

- call/2: This consumes all the requests. This will be responsible for the plug-like behavior of the Goldcrest.ExampleController module.

- greet/2: This consumes requests matching the GET /greet route. This will be the controller action responsible for requests matching that route:

lib/goldcrest/support/example_controller.ex

```
defmodule Goldcrest.ExampleController do
  # Goldcrest.Controller should define all the helper functions
  import Goldcrest.Controller

  def call(conn, action: action) do
    apply(__MODULE__, action, [conn, conn.params])
  end
end
```

We start by importing the Goldcrest.Controller module. We haven't written this yet, but this module will be responsible for defining the helper functions we will use in Goldcrest.ExampleController, such as render/3.

Then, we defined a call/2 function. This is the function that is explicitly invoked from the Goldcrest.ExampleRouter module along with conn as the first argument and :action as a key in the list of options. Therefore, we're matching for the presence of :action. Once called, this function calls the function in the module matching the given action by using apply/3 with **multi-factor authentication (MFA)**.

---

apply/3 **and** __MODULE__

Elixir provides a way to dynamically invoke functions with the module name, function name, or arguments defined at runtime by using the Kernel.apply/3 function. The arguments to the apply/3 function are in the MFA format described in the previous chapters, where the first argument is the module name, the second is the function name, and the third is the list of arguments to be passed to the function. When the module name is the current module, use __MODULE__, which is the shorthand for referring to the current module.

In cases like the ExampleController.call/2 function where we didn't know what action needed to be invoked at compile time, apply/3 is used to add that dynamic behavior.

However, if you do know the function name and the number of arguments at the time of compilation, it is advised to use the module.function(arg1, ..., argN) syntax because it's more explicit. Contrary to popular belief, there isn't a performance benefit to using either because of compiler optimization.

Next, let's define the greet/2 function:

lib/goldcrest/support/example_controller.ex

```
defmodule Goldcrest.ExampleController do
  # ...
  def greet(conn, _params) do
    conn
    |> put_status(200)
    |> render(:json, %{data: "hello world"})
  end
end
```

In the preceding code snippet, we added the greet/2 function to our controller. This function acts as the controller action for the GET /greet route and simply responds with a 200 status and a JSON message.

For our controller to work, we need to define the Goldcrest.Controller module, which was imported into our Goldcrest.ExampleController module, and define helper functions such as put_status/2 and render/3.

Since our controller actions are plugs, which involve transforming the %Plug.Conn{} struct, we can derive a lot of our functionality from the Plug.Conn module.

So, let's define the Goldcrest.Controller module:

lib/goldcrest_controller.ex

```
defmodule Goldcrest.Controller do
  import Plug.Conn

  def render(conn, :json, data) when is_map(data) do
    status = conn.status || 200

    conn
    |> put_resp_content_type("application/json")
    |> send_resp(status, Jason.encode!(data))
  end
end
```

We started by importing the Plug.Conn module, which gives us access to the functions we can use to manipulate the %Plug.Conn{} struct. Then, we define render/3 only for rendering the :json data, while checking that the data variable is a map. In the render/3 function, we simply set the response's content type to application/json and use Plug.Conn.send_resp/3 to send a response with the encoded map as the body. Before sending the response, Plug.Conn.send_resp/3 also checks whether a status is set for the %Plug.Conn{} struct. If it is set, it uses the previously set status; else, it uses 200.

We could similarly define the Goldcrest.put_status/2 function, but Plug.Conn.put_status/2 is already defined and has a simple enough interface. Therefore, instead of defining the put_status/2 function in Goldcrest.Controller, we can simply import Plug.Conn in our example controller module:

lib/goldcrest/support/example_controller.ex

```
defmodule Goldcrest.ExampleController do
  # ..
  import Plug.Conn
  # ..
end
```

Doing this allows us to not only use the put_status/2 function but also gives us access to other useful functions such as put_session/3, get_session/2, and so on, which allows us to write better controller actions. In fact, Phoenix.Controller does exactly this to bring the power of Plug.Conn to all of its controllers.

Now that we have a way to define a controller, let's test it with a web server.

Integration with web servers

We can use Plug.Cowboy in a similar way to *Chapter 3* to build a server that dispatches the GET /greet calls to the ExampleController module. Let's start by creating an iex session with the goldcrest app loaded. We can then require the example router and controller files to use them as part of our HTTP server:

```
$ iex -S mix
iex> Code.require_file("./lib/goldcrest/support/example_controller.
ex")
iex> Code.require_file("./lib/goldcrest/support/example_router.ex")
iex> Mix.install([{:plug_cowboy, "~> 2.0"}])
iex> opts = [
...>    scheme: :http,
...>    plug: Goldcrest.ExampleRouter,
...>    options: [port: 4040]
...> ]
```

```
iex> %{start: {mod, fun, args}} = Plug.Cowboy.child_spec(opts)
iex> apply(mod, fun, args)
{:ok, #PID<_,_,_>}
```

Calling the preceding list of functions from `iex` starts a Cowboy HTTP server on port 4040. We can now use `curl` to test our HTTP server with the controller:

```
$ curl -v http://localhost:4040/greet
*    Trying 127.0.0.1:4040...
* Connected to localhost (127.0.0.1) port 4040 (#0)
> GET /greet/ HTTP/1.1
> Host: localhost:4040
> User-Agent: curl/7.79.1
> Accept: */*
>
* Mark bundle as not supporting multiuse
< HTTP/1.1 200 OK
< cache-control: max-age=0, private, must-revalidate
< content-length: 15
< content-type: application/json; charset=utf-8
< date: Sun, 21 Nov 2021 23:44:29 GMT
< server: Cowboy
<
* Connection #0 to host localhost left intact
{"status":"ok"}%
```

In the preceding response, we can also see that the HTTP status was `200 OK`, as expected.

To kill the Cowboy HTTP server, run the following commands in the same `iex` session:

```
$ iex -S mix
# .... same iex session
{:ok, #PID<_,_,_>}
iex> {:ok, pid} = v()
iex> Process.exit(pid, :kill)
[error] GenServer #PID<_._._> terminating
** (stop) killed
nil
```

Similarly, we can test this with `plug_goldcrest_http_server` to ensure our controller works with our own HTTP server:

```
$ iex -S mix
iex> Code.require_file("./lib/goldcrest/support/example_controller.
ex")
iex> Code.require_file("./lib/goldcrest/support/example_router.ex")
iex> Mix.install([{:plug_goldcrest_http_server, "~> 0.1.0"}])
```

```
iex> opts = [
...>    plug: Goldcrest.ExampleRouter,
...>    port: 4040,
...>    options: []
...> ]
iex> %{start: {mod, fun, args}} = Plug.Goldcrest.HTTPServer.child_
spec(opts)
iex> apply(mod, fun, args)
{:ok, #PID<_,_,_>}


22:42:49.877 [info]  Started a webserver on port 4040
```

Now, by running the `curl` request again, we can see the controller works well with our HTTP server:

```
$ curl -v http://localhost:4040/greet
*   Trying 127.0.0.1:4040...
* Connected to localhost (127.0.0.1) port 4040 (#0)
> GET /greet/ HTTP/1.1
> Host: localhost:4040
> User-Agent: curl/7.79.1
> Accept: */*
>
* Mark bundle as not supporting multiuse
< HTTP/1.1 200
< cache-control: max-age=0, private, must-revalidate
< content-length: 15
< content-type: application/json; charset=utf-8
<
* Excess found in a read: excess = 1, size = 15, maxdownload = 15,
bytecount = 0
* Closing connection 0
{"status":"ok"}%
```

Similar to the Cowboy server's response, we can see that the HTTP status for the response is 200 OK.

Now, we have a fully functioning controller that can handle HTTP requests. One of the key responses of a controller, however, is sending a redirect response, which invokes a redirect action on the browser side. We will see how to implement that in the next section.

Understanding redirects

One of the most used `Phoenix.Controller` functions is `redirect/2`. This is often used at the controller level to redirect `%Plug.Conn{}` to a different route after setting a flash message. The following is an example of a controller action using `redirect/2`:

```
defmodule RedirectExampleController do
  use RedirectExampleWeb, :controller

  def show(conn, show_params) do
    show_info = ShowInfo.get(show_params)

    render(conn, "show.html", show_info: show_info)
  end

  def create(conn, create_params) do
    created = Creator.create(create_params)

    conn
    |> put_flash(:info, "Successfully created!")
    |> redirect(to: Routes.redirect_example_path(conn,
        :show, created))
  end
end
```

In the preceding Phoenix controller, there are two actions: `show` and `create`. The `show` action simply fetches the data and renders an HTML. However, the `create` action runs a command and redirects the request to the `show` action, with a specific parameter. It also sets a `flash` message before redirecting, but we will not cover flash messages at this point.

This is one of the most common uses of the `redirect/2` function. Therefore, our controller interface is incomplete without implementing `redirect/2`.

To implement `redirect/2`, we need to first understand how redirects work.

Redirects work by simply updating the current page's location or window to a different location. In JavaScript, this can be done by updating an `href` location directly or calling `replace` on the location. The following are code examples of redirects in JavaScript:

```
// manually update location
window.location.href = "http://location:4040/redirected_url"

// OR

// update location by calling replace
window.location.replace("http://location:4040/redirected_url")
```

Either way, we need a way to update the browser's location programmatically from the server's response. The best way to signal a browser to change the window's location (redirect) is by sending HTTP codes with a 3XX status, specifically 302, with a location header. This response lets a browser know that the request wants it to perform a URL redirection, and the page being redirected to is Found, which means it's expected to exist.

Here's an example response with a redirect:

```
HTTP/1.1 302 Found
cache-control: max-age=0, private, must-revalidate
content-length: 72
content-type: text/html; charset=utf-8
location: http://location:4040/redirected_url
...

<html>
  <body>
    You are being redirected
      to:
    <a href="/some-url">
      a different site
    </a>
  </body>
</html>
```

Another thing to note here is that adding a body to a 302 response is optional, but often for introspective and debugging purposes, HTTP servers/web frameworks add a body indicating that the request is being redirected.

Now that we know what we need to do to implement redirects, let's add the redirect/2 function to the Goldcrest.Controller module:

lib/goldcrest_controller.ex

```
defmodule Goldcrest.Controller do
  # ..

  def redirect(conn, to: url) do
    body = redirection_body(url)
    status = conn.status || 302

    conn
    |> put_resp_header("location", url)
    |> content_type("text/html")
    |> send_resp(302, body)
```

```
      end

      defp redirection_body(url) do
        html = Plug.HTML.html_escape(url)

        "<html><body>You are being <a href=\"#{html}\">
          redirected</a>" <>
          ".</body></html>"
      end
```

In the preceding code snippet, we have defined a new `redirect/2` function, which takes a `%Plug.Conn{}` struct and `:to` as an option indicating what URL to redirect to.

The `redirect/2` function first generates a response body using the `redirection_body` private function. Note the use of `Plug.HTML.html_escape/1` to avoid any conflicts before injecting `url` as an HTML string.

The generated body is then used as the response body along with the `302` status code, indicating that this is a URL redirection. Before sending the response, it also sets the `location` header of the response to the given `url`.

Now, it's time to test the `redirect/2` function. Let's create a new route in our router:

lib/goldcrest/support/example_router.ex

```
defmodule Goldcrest.ExampleRouter do
  use Plug.Router
  # ..
  get "/greet" do
    Goldcrest.ExampleController.call(conn, action: :greet)
  end

  get "/redirect_greet" do
    Goldcrest.ExampleController.call(conn,
      action: :redirect_greet)
  end
  # ..
end
```

Next, let's define the action, `:redirect_greet`, which the new route will dispatch requests to:

lib/goldcrest/support/example_controller.ex

```
defmodule Goldcrest.ExampleController do
  # ...
  def greet(conn, _params) do
```

```
    conn
    |> put_status(200)
    |> render(:json, %{data: "hello world"})
  end

  def redirect_greet(conn, _params) do
    conn
    |> redirect(to: "/greet")
  end
end
```

In the preceding code snippet, we have added a new action to the ExampleController module, redirect_greet/2, which simply redirects the incoming conn to the GET /greet route.

We can test whether we're getting the expected response with the correct HTML, HTTP code, and location by running a curl request again:

```
$ curl -v http://localhost:4040/redirect_greet
*     Trying 127.0.0.1:4040...
* Connected to localhost (127.0.0.1) port 4040 (#0)
> GET /redirect_greet HTTP/1.1
> Host: localhost:4040
> User-Agent: curl/7.79.1
> Accept: */*
>
* Mark bundle as not supporting multiuse
< HTTP/1.1 302
< cache-control: max-age=0, private, must-revalidate
< content-length: 72
< content-type: text/html; charset=utf-8
< location: /greet
<
* Excess found in a read: excess = 1, size = 72, maxdownload = 72,
bytecount = 0
* Closing connection 0
<html>
  <body>
    Redirecting you to <a href="/greet">this site</a>.
  </body>
</html>
```

We can see that the HTML body is as expected, along with a 302 HTTP response code with the location header pointing to /greet.

We can further test the redirection by visiting http://localhost:4040/redirect_greet from a web browser and expecting to be redirected to http://localhost:4040/greet.

In the following screenshot, we visit `http://localhost:4040/redirect_greet`:

Figure 4.2 – Redirect request

We can see it quickly redirects to `http://localhost:4040/greet`:

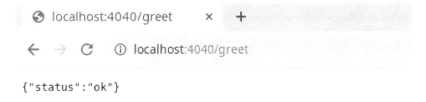

Figure 4.3 – Redirect result

Testing the controller

Now that we have a way to define our controllers, it's time to see how we can test them. Of course, we could simply test the entire HTTP request the same way we tested our HTTP servers in the previous chapters. However, there's a better and simpler way to test the controller responses.

Since our routers and our controllers are plugs, we can test them using the `Plug.Test` module. The `Plug.Test` module contains a collection of helper functions that are designed to help us test plugs better.

Here is how we can use `Plug.Test` to test our controller's `/greet` action:

test/goldcrest/support/example_controller_test.exs

```
defmodule Goldcrest.ExampleControllerTest do
  use ExUnit.Case
  use Plug.Test

  describe "GET /greet" do
    test "responds with 200 status" do
```

```
        conn = conn(:get, "/greet")

        conn = Goldcrest.ExampleRouter.call(conn, [])

        assert conn.status == 200
        assert conn.resp_body == Jason.encode!(%{status:
                                            "ok"})
      end
    end
  end
```

In the preceding test, we see how to test a controller just as a plug without explicitly starting the HTTP server. We are creating a %Plug.Conn{} struct by calling the conn/2 function imported by the Plug.Test module. Then, we're explicitly calling the ExampleRouter plug and capturing the returned conn to analyze the results.

This allows us to write smaller tests that are much quicker to execute because the requests don't have to go through the TCP layer of our application.

Similarly, we can test the /redirect_greet route as follows:

test/goldcrest/support/example_controller_test.exs

```
defmodule Goldcrest.ExampleControllerTest do
  # ...

  describe "GET /redirect_greet" do
    test "responds with a redirect status" do
      conn = conn(:get, "/redirect_greet")

      conn = Goldcrest.ExampleRouter.call(conn, [])

      assert conn.status == 302
      assert conn.resp_body =~ "You are being"
      assert conn.resp_body =~ "redirected"
      assert Plug.Conn.get_resp_header(conn, "location") ==
        ["/greet"]
    end
  end
end
```

In the preceding code snippet, we wrote a test for the /redirect_greet route. The test is quite similar to the one for the /greet route; the differences are that we're testing for the status to equal 302 and expecting the location header to be properly set.

In future chapters, we will develop a cleaner and simpler interface to write these controller tests. For now, we can keep using `Plug.Test` to test our controllers independent of the HTTP server.

Summary

In this chapter, we learned what controllers are both within and outside the context of a simple application. We then learned how Phoenix implements its controllers by utilizing Plug and how the entire request-response cycle, right from the router to the controllers, is powered by Plug. Taking that philosophy, we built our own controller interface and tested an example controller with both `goldcrest_http_server` and `cowboy`. To make our controller functionality more complete, we added the ability to redirect a request to another route by learning how redirects actually work in the context of a web browser. We finally wrapped up by using the `Plug.Test` module to test our controller without having to start our HTTP server.

Similar to previous chapters, we didn't cover all the functions that should be added to the `Goldcrest.Controller` module to save time. We also didn't write the `render/3` implementation for HTML pages because we will cover that in a future chapter where we will talk about views.

Exercises

In the `Goldcrest.Controller` module, we have added the ability to `render` a JSON response and `redirect` a request to a different URL, but there's a lot more we could add to that module.

- How would you go about implementing the `send_file/2` function? This is just a thought exercise since it's not straightforward to implement it.

- What other functions should we add to the `Goldcrest.Controller` module?

5

Adding Controller Plugs and Action Fallback

In the last chapter, we built a framework to write a controller and wrapped it inside the `Goldcrest.Controller` module. In this chapter, we will learn about extending the functionality of our controllers by intercepting an incoming request and calling specific plugs on it based on the controller's definition before letting the controller's actions handle that request. This will give developers the means to add better control of the request flow while keeping their controller code simple and digestible. We will learn how to use the `Plug.Builder` module to build a pipeline of plugs in a controller. We will then learn how to apply a pipeline of plugs to an action and how to define a fallback controller to capture any errors.

By the end of the chapter, you will understand `Plug.Builder` enough to build complex plug pipelines without needing Phoenix or any other libraries other than Plug itself. You will also learn how Phoenix implements the fallback controller using the same philosophy as `Plug.Builder`.

Here are the exact topics we will cover in this chapter:

- Understanding `Plug.Builder`
- Using `Plug.Builder` to extend our controller's functionality
- Testing the `Plug.Builder` pipelines
- Defining a fallback module for controller errors
- Testing fallback controller errors

Technical requirements

This chapter, just like the previous one, is loaded with code snippets. It is recommended that readers follow along with the coding. We will still use Elixir 1.12.x and Erlang 23.2.x, so ensure those are installed on your computer.

I will make the same assumption that you've read through the previous chapters, coding along whenever we reach a snippet. For this chapter, it is particularly important that you've read the previous chapter and understood how we created our own controller. We will be using the code from *Chapter 4* to update the `Goldcrest.Controller` module.

Though not necessary, any experience writing Phoenix and Rails controllers would help you better understand this chapter's contents.

The code examples for this chapter can be found at `https://github.com/PacktPublishing/ Build-Your-Own-Web-Framework-in-Elixir/tree/main/chapter_05`

Plug pipeline in controllers

In many web frameworks, there's a way to filter and intercept requests that are dispatched to a controller's action and perform specific checks on them. For example, Rails has filters, specifically the `before_action` hooks, which are called before the request reaches the controller's action. These checks can log a request, perform a redirect, or even halt the request-response cycle. A well-known usage of these filters is to ensure that an incoming request is authenticated:

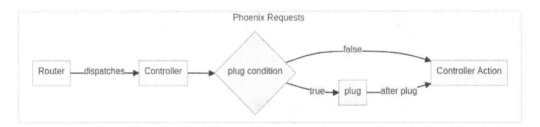

Figure 5.1 – Plug pipeline in Phoenix

Here's how authentication using filters looks in a Rails controller:

```
class PageController < ApplicationController
  before_action :ensure_authenticated!
  before_action :ensure_authorized!, only: [:index]

  def index
    # index action related code
  end

  def show
    # show action related code
  end

  private
```

```
  def ensure_authenticated!
    unless authenticated?
      flash[:error] = "Please Sign-in"
      redirect_to new_sign_in_url
    end
  end

  def ensure_authorized!
    unless admin?
      head :forbidden
    end
  end
end
```

In the preceding code snippet, we see that all actions in `PageController` need to go through a `before_action` call to `ensure_authenticated!`, and only the `index` action needs to go through `ensure_authorized!`. The `before_action` interface makes it very easy to define a list of filters a request must go through before hitting a controller action.

Phoenix has something very similar to `before_action`. Just like a Phoenix router, a Phoenix controller can define its own pipeline. Unlike the routers, though, a controller only has a single pipeline, which can optionally be called for an action or specific guard clause.

Let's take a look at the following Phoenix controller:

```
defmodule ExampleWeb.PageController do
  use ExampleWeb, :controller

  plug :ensure_authenticated!
  plug :ensure_authorized!,
       role: :admin when action == :index

  def index(conn, _) do
    # index action related code
  end

  def show(conn, _) do
    # show action related code
  end

  def ensure_authenticated!(conn, _options) do
    if authenticated(conn) do
      conn
    else
      conn
      |> put_flash(:error, "Please Sign-in")
```

```
            |> redirect(to: "/sign_in")
            |> halt()
      end
    end

    def ensure_authorized!(conn, role: role) do
      if Enum.member?(roles(conn), role) do
        conn
      else
        conn
        |> put_status(403)
        |> halt()
      end
    end
  end
```

Similar to the Rails code snippet, this controller calls `ensure_authenticated!` for all of its actions and `ensure_authorized!` for only the `index/2` action. We are also halting a connection in those plugs if a set of conditions isn't met. A Phoenix controller knows not to call the action if a connection is halted in a previously invoked plug.

Now that we know how controller-level plugs look in Phoenix and Rails, let's learn how to implement in our web framework.

Understanding the Plug.Builder module

To add a similar plug pipeline to our controller, we will need to keep track of all the `plug/3` calls in the controller and apply those plugs in the correct order whenever the controller receives a request. This sounds like a lot of work to do from scratch, but fortunately, the Plug package provides a module that could help – the `Plug.Builder` module.

The `Plug.Builder` module can be used to define a pipeline of plugs that are executed one after the other. It defines a `plug/2` macro, which is the main interface to build a pipeline by adding a plug to the list of already registered plugs. The plugs in the pipeline are executed in the order of their definition, and the connection returned by the final plug is then passed along to the main plug call.

Here's an example of this:

```
defmodule ExamplePlugPipeline do
  use Plug.Builder

  plug :plug1
  plug :plug2

  def plug1(conn, _opts) do
```

```
      conn
      |> put_session(:plug1, true)
    end

  def plug2(conn, _opts) do
    conn
    |> put_session(:plug2, true)
    |> send_resp(200, "Done!")
    end
  end
```

As you can see from the preceding module, calling use with `Plug.Builder` also imports the `Plug.Conn` module. This gives us access to functions such as `put_session/3` and `send_resp/3`, further helping us write cleaner plugs for the pipeline.

The act of using `Plug.Builder` makes a module a plug. This is because it defines a set of functions and behavior inside the module, which allows it to be invoked just like a `Plug` module.

`Plug.Builder` can be used with the following options:

- `log_on_halt`: This allows us to configure what log level to use when a pipeline is halted.

- `init_mode`: This allows us to configure when to initialize the plug options. This can either be set to `:compile` or `:runtime`. It defaults to `:compile`, but an example of using this at `:runtime` would be when using a runtime configuration to authorize a request.

How is `Plug.Builder` implemented?

The `Plug.Builder` module itself is a plug under the hood. It implements the Plug behavior. Therefore, it has implementations defined for `init/2` and `call/2`. This makes it extremely easy to integrate `Plug.Builder` with any HTTP server or a pre-existing Plug application.

`Plug.Builder` stores the list of "registered" plugs added to the pipeline by calling the `plug/2` macro as a module attribute: `@plugs`. Then, when it's finally invoked using the `call/2` function, it uses the `@plugs` attribute to get a list of registered plugs and invokes them in the correct order.

As part of the `call/2` function, `Plug.Builder` checks whether the connection struct that was returned from the previous plug is halted or not. If it is halted, it returns the connection without calling the next plug. Otherwise, it proceeds with the rest of the plug pipeline.

The `Plug` API functions implemented in the `Plug.Builder` module can also be overridden to further extend the module's behavior.

However, overriding the `call/2` function means overriding the `Plug.Builder` module's functionality, which sequentially invokes the list of plugs defined by the `plug/2` macro. Therefore, generally, when the `Plug.Builder` module's `call/2` is overridden, it is accompanied by manually invoking its original implementation to access the ability to sequentially call the defined plug pipeline.

Let's look at the following code snippet:

```
defmodule ExamplePlugPipeline do
  use Plug.Builder

  plug :plug1
  plug :plug2

  def call(conn, opts) do
    conn
    |> super(opts) # calls the original implementation
    |> assign(:called_plugs, @plugs)
  end

  # ..
end
```

In the preceding module, we overrode the `call/2` function and simply added a call to the `assign/3` function, which sets the connection's `assigns` value for the `:called_plugs` key to the `@plugs` module attribute. At the time of compilation, the module attribute contains a list of all the plugs registered until that point in the file. Therefore, this is a quick way of adding a list of plugs to the connection after they are called.

`defoverridable` **and** `super` **in Elixir**

In Elixir, we can use `defoverridable` to make a function overridable outside the current module. This is possible because of macros. Once a function is marked as overridable, Elixir knows to define that function lazily, allowing developers to override it in the context of the same module later in the module definition.

Elixir also provides a way to call the original implementation of an overridable function while it's being overridden. This is done by using the `super/1` macro defined in the `Kernel.SpecialForms` module. This allows the developers to override a function while still keeping some of its functionality.

Elixir also allows developers to override a list of callbacks defined using a "behavior" by simply defining the "behavior" module as overridable.

Once a plug returns a halted connection, `Plug.Builder` prevents any further plugs in the pipeline from being called and therefore returns the halted connection.

For example, in the following code snippet, we have a pipeline of two plugs:

```
defmodule HaltPlugPipeline do
  use Plug.Builder

  plug :plug1
  plug :halt
  plug :plug2

  def plug1(conn, _) do
    conn
    |> put_session(:plug1_called, true)
  end

  def halt(conn, _) do
    conn
    |> halt()
  end

  def plug2(conn, _) do
    conn
    |> put_session(:plug2_called, true)
  end
end
```

In the preceding plug pipeline, `plug2` will never be invoked because `halt` always returns a halted connection. Therefore, `Plug.Builder` will prevent `:plug2` from being called.

Let's test this assertion by using the `Plug.Test` module, as shown in the following code snippet:

```
defmodule HaltPlugPipelineTest do
  use ExUnit.Case
  use Plug.Test

  describe "call/2" do
    test "invokes only plug1 and returns a halted
          connection" do
      conn = conn(:get, "/")

      conn = HaltPlugPipeline.call(conn, [])

      # Returns a halted connection
      assert conn.halted

      # Calls plug1
      assert get_session(conn, :plug1_called)
```

```
        # Doesn't call plug2
        refute get_session(conn, :plug2_called)
      end
    end
end
```

In the preceding test, we're invoking the `HaltPlugPipeline` plug defined by calling `use` with the `Plug.Builder` module. We can see that the test returns a halted connection due to a call to the `halt/2` function. We also test that the returned connection has a session for `:plug1_called` set to `true` but not `:plug2_called`, implying that the pipeline didn't proceed after the `:halt` plug call.

Now that we have learned how the `Plug.Builder` module works, we can use that to implement our own controller pipeline.

Building the controller pipeline

Let's start with the example controller we created in the previous chapter as part of an example for the `Goldcrest.Controller` module:

---

lib/goldcrest/support/example_controller.ex

```
defmodule Goldcrest.ExampleController do
  import Plug.Conn
  import Goldcrest.Controller

  def call(conn, action: action) do
    apply(__MODULE__, action, [conn, conn.params])
  end

  def greet(conn, _params) do
    conn
    |> put_status(200)
    |> render(:json, %{status: "ok"})
  end

  # ..
end
```

And just as in the previous chapter, we can start an HTTP server serving the preceding controller by running an `iex` shell and loading the controller and router files:

```
$ iex -S mix
iex> Code.require_file("./lib/goldcrest/support/example_controller.
ex")
```

```
iex> Code.require_file("./lib/goldcrest/support/example_router.ex")
iex> Mix.install([{:plug_cowboy, "~> 2.0"}])
iex> opts = [
...>    scheme: :http,
...>    plug: Goldcrest.ExampleRouter,
...>    options: [port: 4040]
...> ]
iex> %{start: {mod, fun, args}} = Plug.Cowboy.child_spec(opts)
iex> apply(mod, fun, args)
{:ok, #PID<_,_,_>}
```

We can then ensure the server is working by making a `curl` request:

```
$ curl -v http://localhost:4040/greet
*   Trying 127.0.0.1:4040...
* Connected to localhost (127.0.0.1) port 4040 (#0)
> GET /greet/ HTTP/1.1
> Accept: */*
>
* Mark bundle as not supporting multiuse
< HTTP/1.1 200 OK
< cache-control: max-age=0, private, must-revalidate
< content-length: 15
< content-type: application/json; charset=utf-8
< server: Cowboy
<
{"status":"ok"}%
```

Now, let's add the `Plug.Builder` module to our controller and define a plug pipeline that checks for an authorization header in an incoming request:

lib/goldcrest/support/example_controller.ex

```
defmodule Goldcrest.ExampleController do
  use Plug.Builder
  import Plug.Conn
  import Goldcrest.Controller

  def call(conn, action: action) do
    conn = super(conn, [])

    unless conn.state == :sent or conn.halted do
      apply(__MODULE__, action, [conn, conn.params])
    else
      conn
```

```
        end
    end

    # ..
  end
```

In the preceding code snippet, we added a use call to the `Plug.Builder` module. As we learned earlier in this chapter, `Plug.Builder` already defines an overridable `call/2` function. Therefore, we used the `super/2` macro to invoke `Plug.Builder.call/2` (the original implementation) to call any plugs defined in the pipeline. We then check the state of the returned connection, and if it's not `:sent` or `halted`, we call the controller action.

Now, we can add a plug, `ensure_authorized!`, which checks for an authorization header in the incoming connection:

lib/goldcrest/support/example_controller.ex

```elixir
defmodule Goldcrest.ExampleController do
  use Plug.Builder
  import Goldcrest.Controller

  plug :ensure_authorized!

  # ..

  defp ensure_authorized!(conn, _opts) do
    if authorized?(conn) do
      conn
    else
      conn
      |> put_status(401)
      |> render(:json, %{status: "Unauthorized"})
      |> halt()
    end
  end

  defp authorized?(conn) do
    auth_header = get_req_header(conn, "authorization")

    auth_header == ["Bearer #{token()}"]
  end

  defp token, do: "secret"
end
```

In the preceding code snippet, we used the `Plug.Conn` functions imported from calling `use` with the `Plug.Builder` module to check whether the request header contains the correct authorization token. For the sake of simplicity, we're setting the token to a compile-time value of `"secret"`.

Let's restart our server:

```
$ iex -S mix
iex> Code.require_file("./lib/support/example_controller.ex")
iex> Code.require_file("./lib/support/example_router.ex")
iex> Mix.install([{:plug_cowboy, "~> 2.0"}])
iex> opts = [
...>    scheme: :http,
...>    plug: Goldcrest.ExampleRouter,
...>    options: [port: 4040]
...> ]
iex> %{start: {mod, fun, args}} = Plug.Cowboy.child_spec(opts)
iex> apply(mod, fun, args)
{:ok, #PID<_,_,_>}
```

We can now see that making a request without an authorization header returns a 401 response:

```
$ curl -v http://localhost:4040/greet
*    Trying 127.0.0.1:4040...
* Connected to localhost (127.0.0.1) port 4040 (#0)
> GET /greet HTTP/1.1
> Host: localhost:4040
> User-Agent: curl/7.79.1
> Accept: */*
>
* Mark bundle as not supporting multiuse
< HTTP/1.1 200 OK
< cache-control: max-age=0, private, must-revalidate
< content-length: 25
< content-type: application/json; charset=utf-8
< server: Cowboy
<
* Connection #0 to host localhost left intact
{"status":"Unauthorized"}%
```

If we pass the wrong authorization header, it still returns a 401 response:

```
$ curl --header "Authorization: Bearer badtoken" -v http://
localhost:4040/greet
*    Trying 127.0.0.1:4040...
* Connected to localhost (127.0.0.1) port 4040 (#0)
> GET /greet HTTP/1.1
> Host: localhost:4040
```

```
> User-Agent: curl/7.79.1
> Accept: */*
> Authorization: Bearer badtoken
>
* Mark bundle as not supporting multiuse
< HTTP/1.1 200 OK
< cache-control: max-age=0, private, must-revalidate
< content-length: 25
< content-type: application/json; charset=utf-8
< server: Cowboy
<
* Connection #0 to host localhost left intact
{"status":"Unauthorized"}%
```

If, however, we pass the correct authorization header, we get a 200 (successful) response:

```
$ curl --header "Authorization: Bearer secret" -v http://
localhost:4040/greet
*     Trying 127.0.0.1:4040...
* Connected to localhost (127.0.0.1) port 4040 (#0)
> GET /greet/ HTTP/1.1
> Host: localhost:4040
> User-Agent: curl/7.79.1
> Accept: */*
> Authorization: Bearer secret
>
* Mark bundle as not supporting multiuse
< HTTP/1.1 200 OK
< cache-control: max-age=0, private, must-revalidate
< content-length: 15
< content-type: application/json; charset=utf-8
< server: Cowboy
<
* Connection #0 to host localhost left intact
{"status":"ok"}%
```

Now, we can define a plug pipeline for our controller. Unlike Phoenix, however, our controller cannot define plugs that are conditionally invoked based on a guard clause. We will work on adding this functionality once we've covered metaprogramming in the next section.

Implementing action fallback

Phoenix supports using a plug as a fallback to any controller action. This plug can be used to translate data returned by a controller action that is not a %Plug.Conn{} struct into a %Plug.Conn{} struct. The fallback plug is registered by calling the Phoenix.Controller.action_fallback/1 macro.

The following is a controller that uses action_fallback:

```
defmodule ExampleWeb.PageController do
  use Phoenix.Controller

  action_fallback ExampleWeb.FallbackController

  def index(conn, params) do
    case list_pages(params) do
      {:ok, pages} -> render(conn, "index.html",
                            pages: pages)
      other -> other # includes {:error, :bad_params}
    end
  end

  # ..
end
```

We can see in the preceding controller that if list_pages/1 doesn't return {:ok, pages}, the action itself doesn't return a connection. Phoenix, very smartly, delegates anything other than a %Plug.Conn{} struct to the module registered as action_fallback, along with the original connection that was passed to the action. In this case, if list_pages/1 returns {:error, :bad_params}, it calls ExampleWeb.FallbackController.call(conn, {:error, :bad_params}). An advantage of doing something like this is that it allows developers to centralize error handling from all the controllers to one plug while focusing on the happy paths in the controller itself.

Let's now look at an example of a fallback plug:

```
defmodule ExampleWeb.FallbackController do
  use Phoenix.Controller

  def call(conn, {:error, :bad_params}) do
    conn
    |> put_status(500)
    |> put_view(ExampleWeb.ErrorView)
    |> render("500.html")
  end
end
```

In the preceding plug, we define a call function that knows how to handle any unhandled controller actions that return {:error, :bad_params}. In this case, it simply updates the status to 500 and renders an error page.

Now, that we've seen action_fallback in action, let's try implementing it for our controller. We can start by updating an action, redirect_greet/2, to conditionally return something other than %Plug.Conn{}, preferably an error tuple ({:error, term()}):

```
defmodule Goldcrest.Controller.ExampleController do
  # ..
  def redirect_greet(conn, params) do
    case validate_params(params) do
      :ok ->
        conn
        |> redirect(to: "/greet")

      other ->
        other
    end
  end

  defp validate_params(%{"greet" => "true"}), do: :ok
  defp validate_params(_), do: {:error, :bad_params}

  # ..
end
```

After making the preceding change, redirect_greet/2 only redirects if it is called with the %{"greet" => "true"} parameters. Otherwise, it delegates the error to a fallback module, which we will work on next.

Now, let's update the call/2 function to invoke a registered fallback plug if an action doesn't return the %Plug.Conn{} struct. We can do that by first extracting out the apply/3 call into a simpler function called apply_action/2, and then updating apply_action/2 to check for the return value when an action is invoked:

```
defmodule Goldcrest.Controller.ExampleController do
  # ..
  @fallback_plug Goldcrest.Controller.ExampleFallback

  def call(conn, action: action) do
    conn = super(conn, [])

    unless conn.state == :sent or conn.halted do
      apply_action(conn, action)
    else
```

```
      conn
    end
  end

  defp apply_action(conn, action) do
    case apply(__MODULE__, action, [conn, conn.params]) do
      %Plug.Conn{} = conn -> conn
      other -> @fallback_plug.call(conn, other)
    end
  end

  # ..
end
```

In the preceding code snippet, we have defined `fallback_plug` as a module attribute, `@fallback_plug`, which gets called whenever an action doesn't return a connection.

Now, let's go ahead and implement the `Goldcrest.Controller.ExampleFallback` plug:

```
defmodule Goldcrest.Controller.ExampleFallback do
  import Plug.Conn
  import Goldcrest.Controller

  def call(conn, {:error, :bad_params}) do
    conn
    |> put_status(500)
    |> render(:json, %{error: "bad params"})
  end
end
```

In the preceding module, we defined a `call/2` function similar to the `call/2` function that was defined in the Phoenix callback module earlier. This function knows how to handle the `{:error, :bad_params}` response from a controller action.

For this to work, we need to test out our controller action with some query parameters. We just need one more plug to translate the URL's query parameters into the parameters assigned to the connection. Let's add a `Plug.Parsers` call to our router, which will parse the query string and add it to our connection for further use by the controller:

```
defmodule Goldcrest.ExampleRouter do
  use Plug.Router

  # Add this plug
  plug Plug.Parsers,
    parsers: [:urlencoded, :json],
    json_decoder: Jason
```

```
    # ..
end
```

Now, we're ready to test out our controller with the fallback plug. Let's restart our server and make a `curl` request to the `redirect_greet` path without the parameter first:

```
$ curl --header "Authorization: Bearer secret" -v \
  "http://localhost:4040/redirect_greet"
*   Trying 127.0.0.1:4040...
* Connected to localhost (127.0.0.1) port 4040 (#0)
> GET /redirect_greet HTTP/1.1
> Host: localhost:4040
> User-Agent: curl/7.79.1
> Accept: */*
> Authorization: Bearer secret
>
* Mark bundle as not supporting multiuse
< HTTP/1.1 200 OK
< cache-control: max-age=0, private, must-revalidate
< content-length: 22
< content-type: application/json; charset=utf-8
< server: Cowboy
<
* Connection #0 to host localhost left intact
{"error":"bad params"}%
```

We can see that the preceding response was `{"error": "bad params"}`. Therefore, our fallback is working as expected.

Now, to make sure `redirect_greet` is also working, let's make a final request with the proper parameters:

```
$ curl --header "Authorization: Bearer secret" -v \
  "http://localhost:4040/redirect_greet?greet=true"
*   Trying 127.0.0.1:4040...
* Connected to localhost (127.0.0.1) port 4040 (#0)
> GET /redirect_greet?greet=true HTTP/1.1
> Host: localhost:4040
> User-Agent: curl/7.79.1
> Accept: */*
> Authorization: Bearer secret
>
* Mark bundle as not supporting multiuse
< HTTP/1.1 302 Found
< cache-control: max-age=0, private, must-revalidate
```

```
< content-length: 72
< content-type: text/html; charset=utf-8
< location: /greet
< server: Cowboy
<
* Connection #0 to host localhost left intact
<html><body>You are being <a href="/greet">redirected</a>.</body></
html>%
```

As expected, we got a redirected response. Therefore, we can conclude that the redirect_greet/2 action works as expected with our new fallback module.

Summary

In this chapter, we learned how modern web frameworks such as Rails and Phoenix define a way to hook into a request from a controller, check for certain conditions, and in some cases, halt the request pipeline. We then learned about the details of the Plug.Builder module, how to implement it, and how we can use it to add that feature into our controller.

After learning how to test our Plug.Builder module, we learned how to use the overridable call/2 function to define a fallback plug for a controller action. In this way, we can standardize the error handling across many controllers, allowing developers to focus on the happy path while writing the controllers.

Now that we have a working set of controllers and routers for our web framework, we will focus on adding views and HTML templates in a dynamic, server-rendered manner in the next chapter.

6

Working with HTML and Embedded Elixir

In the previous chapter, we learned how to leverage plugs to implement controllers and routers for our web framework. We were able to receive an HTTP request from our web server and, using our controllers, respond with HTML content. However, the HTML content was always static. In order to support server-side rendering of HTML in a more dynamic way, we need the ability to define HTML templates that have parts that can be dynamically evaluated right before responding. Rails does this using HTML files embedded with Ruby code, which is dynamically evaluated right before the response. Phoenix, before version 1.6, used a similar strategy using the EEx package.

In this chapter, we will dig into the EEx library and learn how to use its API. We will learn how it works and how it can be used to add templating abilities to our server-side rendered HTML responses.

In this chapter, we will cover the following topics:

- What is EEx?
- How to use EEx
- Integrating with EEx to send dynamic HTML content
- Testing dynamically generated HTML conventionally
- Using Floki, an HTML parser, to test dynamically generated HTML

Technical requirements

This chapter relies on Elixir 1.12.x and Erlang 23.2.x. The code snippets in this chapter are relatively simple, so if you're short on time, this would be a good chapter to not code along with. I still recommend following along, especially if you're new to Elixir.

Also, similar to previous chapters, we will be using curl to test the responses from our HTTP server. Therefore, a terminal multiplexer would again be helpful.

The code examples for this chapter can be found at `https://github.com/PacktPublishing/Build-Your-Own-Web-Framework-in-Elixir/tree/main/chapter_06`

Understanding and exploring embedded elixir

Embedded Elixir allows us to embed Elixir code inside a string in a consistent and robust way, outlined by the EEx package. This further allows us to write templates that can be dynamically evaluated using Elixir.

EEx makes the following code possible:

```
$ iex
iex> string = "Hello <%= @person %>"
"Hello <%= @person %>"
iex> EEx.eval_string(string, assigns: [person: "World"])
"Hello World"
```

In the preceding code snippet, we were able to add the ability to evaluate an otherwise static string dynamically using Elixir. In the string, anything inside the <% and %> tags is evaluated as Elixir. Therefore, if we have the variables or functions in the context of evaluation, inside those tags they will get evaluated on calling `EEx.eval_string/3`. Here's an example where we evaluate a function inside a string using embedded Elixir:

```
$ iex
iex> string = "Hello <%= String.downcase(@person) %>"
"Hello <%= String.downcase(@person) %>"
iex> EEx.eval_string(string, assigns: [person: "World"])
"Hello world"
```

Since we called `String.downcase/1` inside the tags this time, it downcased the given string, `"World"`, during evaluation.

We can also use EEx to evaluate the contents of a file. Let's create a new file, `message.eex`:

./message.eex

```
Hello <%= String.capitalize(@person) %>
```

Now, let's use the `message.eex` file as a template to generate a string:

```
$ iex
iex> EEx.eval_file("message.eex", assigns: [person: "world"])
"Hello World"
```

In the preceding example, we can see that calling `EEx.eval_file/3` returns the string with a capitalized `World`, since we're calling `String.capitalize/1` inside the eex tags.

Defining functions dynamically in a module

Just as we used `EEx.eval_file/3` to evaluate the contents of a file dynamically, we can also use `EEx.function_from_file/5` to dynamically define functions in a module at compile time.

Let's define the contents of a function in a file:

./function_body.eex

```
<% IO.puts "This function was called with #{a} and #{b}" %>
<%= a + b %>
```

In the preceding file, we do two operations. First, we print (for logging purposes) what the function is called with. Since we wrapped the `IO.puts/1` call in `<%` without `=`, it will not be part of the final string, which is the result of the evaluation of the file.

Now, let's use the preceding file to define a function in a new module:

./module_a.ex

```
defmodule ModuleA do
  require EEx

  EEx.function_from_file(:def, :fun_a, "function_body.eex",
                         [:a, :b])
end
```

In the preceding module, we use `EEx.function_from_file/5` to indicate that we're defining a public function named `fun_a`, which takes two arguments, a and b, and returns the string that is the result of evaluating `function_body.eex` with the passed values for a and b.

We can test this function out in `iex`:

```
$ iex
iex> Code.require_file("module_a.ex")
[
  {ModuleA,
   <<70, 79, 82, 49, 0, 0, 7, 240, 66, 69, 65, 77, 65, 116,
     85, 56, 0, 0, 0, 200, 0, 0, 0, 21, 14, 69, 108, 105,
     120, 105, 114, 46, 77, 111, 100, 117, 108, 101, 65, 8,
     95, 95, 105, 110, 102, 111, 95, ...>>}
]
iex> return_value = ModuleA.fun_a(1, 2)
```

```
This function was called with 1 and 2
"\n3\n"
iex> return_value
"\n3\n"
```

As expected, calling `ModuleA.fun_a/2` with two arguments first prints a message and then sets the stringified version of the sum of two arguments as the return value.

Now that we know how to compile the contents of `eex` as a function, let's take a look at compiling the template as part of the module's AST itself.

Compiling a template with a module

EEx also provides a way to use a template file at compile time to generate an Elixir **Abstract Syntax Tree (AST)**. We will learn more about what ASTs are and how they work in Elixir in *Chapter 7*, but to summarize, it is a tree representation of Elixir code.

Let's start by defining another embedded file, `ast.eex`:

./ast.eex

```
"<%= a %> <%= b %>"
```

The preceding code returns a concatenated string, consisting of two embedded variables, a and b.

Now, let's compile the preceding file using `EEx.compile_file/2` in an `iex` shell:

```
$ iex
iex> EEx.compile_file("ast.eex")
{:__block__, [],
 [
   {:=, [],
    [
      {:arg0, [], EEx.Engine},
      {{:., [],
        [{:__aliases__, [alias: false], [:String, :Chars]},
         :to_string]}, [],
       [{:a, [line: 1], nil}]}
    ]},
   {:=, [],
    [
      {:arg1, [], EEx.Engine},
      {{:., [],
        [{:__aliases__, [alias: false], [:String, :Chars]},
         :to_string]}, [],
       [{:b, [line: 1], nil}]}
```

```
    ]},
    {:<<>>, [],
      [
        "\"",
        {:"::", [], [{:arg0, [], EEx.Engine}, {:binary, [],
          EEx.Engine}]},
        " ",
        {:"::", [], [{:arg1, [], EEx.Engine}, {:binary, [],
          EEx.Engine}]},
        "\"\n"
      ]}
  ]}
```

We can see in the preceding output that calling EEx.compile_file/2 on ast.eex returns a three-element nested tuple. That tuple is the Elixir version of the AST we will learn more about in *Chapter 7*. Let's see what happens after we evaluate that AST:

```
$ iex
iex> ast = EEx.compile_file("ast.eex")
# output hidden
iex> {output, _} = Code.eval_quoted(ast, a: "Hello", b: "World")
iex> output
"\"Hello World\"\n"
```

As we can see in the preceding code, calling Code.eval_quoted/2 on the AST returned by EEx.compile_file/2, with a as Hello and b as World, returned the Hello World output.

Working with EEx options

All functions in the EEx module support several options. We will be covering some of the important ones in this section and how they affect the final output of the evaluation:

- :file: This defaults to "nofile" when a string is being compiled, or the template file path where eex is defined. This can also be used for debugging purposes.

- :line: Similar to :file, this defaults to the line on the template file or 1 if being compiled from a string.

- :trim: This defaults to false, but it can be set to true when we want to remove white space around the evaluated string. This ensures that at least one newline is retained.

- :indentation: This defaults to 0, but it can be set to a positive integer when an indentation is desired for the final string.

- :engine: EEx Engine is used to evaluate the template. This defaults to EEx.SmartEngine.

Using EEx Smart Engine

EEx comes with the ability to override the **engine** that is being used to evaluate a template. The default engine used by EEx is EEx.SmartEngine. This module is built on top of the regular EEx.Engine but provides a way to speed up evaluation of a template by adding a special type of binding, assigns.

Let's first take a look at an example:

./assigns.eex

```
<%= @a %> <%= @b %>
```

Now, let's use the preceding template file to define a function in a module:

./module_b.ex

```
defmodule ModuleB do
  require EEx

  EEx.function_from_file(:def, :fun_b, "assigns.eex", [:assigns])
end
```

Updating our template and function to only take :assigns allows us to send different assigns without having to re-compile the template.

Now, let's test out ModuleB.fun_b/1:

```
$ iex
iex> Code.require_file("module_b.ex")
[
  {ModuleB,
    <<70, 79, 82, 49, 0, 0, 7, 68, 66, 69, 65, 77, 65, 116,
      85, 56, 0, 0, 0,218, 0, 0, 0, 22, 14, 69, 108, 105,
      120, 105, 114, 46, 77, 111, 100, 117, 108, 101, 66,
      8, 95, 95, 105, 110, 102, 111, 95, ...>>}
]
iex> ModuleB.fun_b(a: "Hello", b: "World")
"Hello World\n"
iex> ModuleB.fun_b(a: "Hola", b: "World")
"Hola World\n"
```

This is super helpful to speed up the evaluation of templates while keeping the interface to use the templates simple and consistent. Phoenix uses a similar strategy to evaluate its templates.

Understanding custom markers

EEx supports many **markers** that can be used alongside `<%` tags. An example of those markers is `=`. When the tag is used with `=` as `<%=`, it lets the default engine know that the result of the statement must be added to the evaluated string.

Similar to `=`, EEx supports other markers that can be used when writing a custom engine:

- `" "`: No marker. In the default engine, this implies that the statement needs to be part of the AST, but it shouldn't be part of the final binary that will be returned as the final result.

- `"="`: As explained earlier, this lets the default engine know to evaluate and return the statement as part of the final result.

- `"|"`: No default implementation. This raises a syntax error.

- `"/"`: No default implementation. This raises a syntax error.

The aforementioned markers can be used to write a custom EEx engine that implements EEx. Engine behavior, which we will discuss in the next section.

Building a custom EEx engine

Now, that we've seen how the default `EEx.SmartEngine` works, it's time for us to write a custom engine of our own. As part of this engine, we will be extending the functionality of `EEx.SmartEngine` by adding a new implementation for the `|` marker, and delegating the rest to `EEx.SmartEngine` itself.

Now, let's define a custom EEx engine that behaves like the `EEx.Engine` module. Let's also start by delegating all the functions to `EEx.SmartEngine`:

./custom_engine.ex

```
defmodule CustomEngine do
  @behaviour EEx.Engine

  @delegate_to EEx.SmartEngine

  @impl true
  defdelegate init(opts), to: @delegate_to

  @impl true
  defdelegate handle_body(state), to: @delegate_to

  @impl true
  defdelegate handle_begin(state), to: @delegate_to

  @impl true
```

```
    defdelegate handle_end(state), to: @delegate_to

    @impl true
    defdelegate handle_text(state, meta, text),
      to: @delegate_to

    @impl true
    def handle_expr(state, marker, expr), to: @delegate_to
  end
```

In the preceding module, we have defined a custom engine module that simply delegates all its behavior to EEx.SmartEngine.

Now, let's override the handle_expr/3 function only for the " | " marker. For the purposes of this example, let's say the desired behavior for this marker is calling IO.inspect/2 on the result of the expression. So, let's add two clauses for handle_expr/3. The first one matches on the " | " marker, and the second one is the general clause that delegates to EEx.SmartEngine.

Let's make changes to the CustomEngine module:

./custom_engine.ex

```
defmodule CustomEngine do
  # ..

  @impl true
  def handle_expr(state, "|", expr) do
    expr =
      quote do
        IO.inspect(unquote(expr))
      end

    handle_expr(state, "", expr)
  end

  # General clause
  def handle_expr(state, marker, expr) do
    EEx.SmartEngine.handle_expr(state, marker, expr)
  end
end
```

In the preceding code, we have defined a general clause for `handle_expr/3`, which simply delegates to `EEx.SmartEngine`, along with a clause that matches on " | " as the marker. In the matching clause, we simply update the expression to call `IO.inspect/2` on the given `expr`, and then call `handle_expr/3` with the given state, the updated expression, and " " as the marker, indicating that we don't want the final string to be part of the result.

Now, it's time to test our custom engine module. Let's start by defining a new template:

./custom_ast.eex

```
<%| a + b %>
<%= a %> <%= b %>
```

In the preceding template, we first call `<%|` on the a + b expression, indicating that we want to inspect the output of a + b. We then print the result of concatenating a and b. Therefore, the final result must only contain the concatenation, but evaluating the template should print/inspect a + b.

Now, let's evaluate the preceding template. Let's first use `EEx.SmartEngine` to evaluate it to make sure it results in `EEx.SyntaxError`:

```
$ iex
iex> EEx.eval_file("custom_ast.eex", [a: 1, b: 2])
** (EEx.SyntaxError) ::: unsupported EEx syntax <%| %>
    (the syntax is valid but not supported by the current
      EEx engine)
    lib/eex/engine.ex:211: EEx.Engine.handle_expr/3
    lib/eex.ex:275: EEx.eval_file/3
```

As we can see in the preceding code, the " | " marker isn't implemented by default. Therefore, let's now evaluate it with `CustomEngine` as engine. For that, we'll first need to require the file, `custom_engine.ex`, and then call `EEx.eval_file/3` with `CustomEngine` as the option for engine:

```
$ iex
iex> Code.require_file("custom_engine.ex")
[
  {CustomEngine,
    <<70, 79, 82, 49, 0, 0, 10, 152, 66, 69, 65, 77, 65,
      116, 85, 56, 0, 0, 0, 244, 0, 0, 0, 21, 19, 69, 108,
      105, 120, 105, 114, 46, 67, 117, 115, 116, 111, 109,
      69, 110, 103, 105, 110, 101, 8, 95, 95, ...>>}
]
iex> return_val = EEx.eval_file(
...>   "custom_ast.eex",
...>   [a: 1, b: 2],
...>   engine: CustomEngine
...> )
```

```
3
"\n1 2\n"
iex> return_val
"\n1 2\n"
```

As we can see, only the stringified result of the concatenation of a and b is returned by the expression, and the result of a + b is printed/inspected on evaluation, as expected.

This was one simple application of a custom EEx engine. Next, let's learn how to update our HTTP server to respond with HTML rendered using EEx templates.

Responding with html.eex files

Now that we know how to use EEx to create templates using Elixir, it's time to start responding with embedded Elixir with our HTTP server.

Let's start by creating a new mix project:

```
$ mix new example_server_html_eex
```

In order to use Goldcrest with this new app, let's start by adding goldcrest as a mix dependency in example_server_html_eex:

---

mix.exs

```
defmodule ExampleServerHtmlEex.MixProject do
  # ..

  defp deps do
    [
      {:goldcrest, "~> 0.0.1"}
    ]
  end
end
```

We can fetch the dependency by running $ mix deps.get, and once fetched, let's add a Router module. This router will be very similar to the one we created in *Chapter 5*, with the only difference being the content type of the response. For this router, we will respond with the text/html content:

---

lib/example_server_html_eex/router.ex

```
defmodule ExampleServerHtmlEex.Router do
  use Plug.Router

  alias ExampleServerHtmlEex.Controller
```

```
plug Plug.Parsers,
  parsers: [:urlencoded, :multipart],
  pass: ["text/html", "application/*"]

plug :match
plug :dispatch

get "/greet", do: Controller.call(conn, action: :greet)

match _ do
  send_resp(conn, 404, "<h1>Not Found</h1>")
end
end
```

In the preceding router, we use the Plug.Router plug with the parsers, allowing requests text/html content type. We also use match and the dispatch plug explained in *Chapter 5* to control the flow of the request. Finally, we define a GET route on the /greet path, which routes the request to the ExampleServerHtmlEex.Controller module's greet action, and return a 404 response for any other matches.

Now, it's time to define ExampleServerHtmlEex.Controller with the :greet action, which returns the text/html content that points to a file. This controller will be similar to the one we wrote in *Chapter 5*, which uses Plug.Builder and imports Plug.Conn and Goldcrest. Controller to get useful functions and behaviors to build a controller:

lib/example_server_html_eex/controller.ex

```
defmodule ExampleServerHtmlEex.Controller do
  use Plug.Builder

  import Plug.Conn
  import Goldcrest.Controller

  def call(conn, action: action) do
    conn = super(conn, [])

    apply(__MODULE__, action, [conn, conn.params])
  end

  def greet(conn, _params) do
    conn
    |> put_status(200)
    |> render(:html, "<h1>Hello World</h1>")
```

```
    end
  end
```

The preceding controller defines a `call/2` function similar to *Chapter 5*. It also defines a `greet/2` function that simply returns a `200` response with static HTML content, `Hello World`.

Let's spin up `iex` and start a server to test this controller:

```
$ iex -S mix
iex> opts = [
...>    scheme: :http,
...>    plug: ExampleServerHtmlEex.Router,
...>    options: [port: 4040]
...> ]
iex> %{start: {mod, fun, args}} = Plug.Cowboy.child_spec(opts)
iex> apply(mod, fun, args)
{:ok, #PID<0.313.0>}
```

Now, without quitting the preceding session, visit `http://localhost:4040/greet` on your web browser. Doing this should yield the following result.

Hello World!

Figure 6.1: The greet response

Now that we have a controller set up to respond with static `text/html` content, it's time to use the power of EEx to make our HTML content dynamic.

Let's start by defining our HTML template in the `priv/templates` folder:

priv/templates/greet.html.eex

```
<h1><%= @greeting %> World!</h1>
```

In the preceding template file, we have the final HTML dynamic by making `@greeting` a variable. By using @ variables in the template, we also imply that we will be using the `EEx.SmartEngine` `:assigns` feature, described earlier in the chapter.

Now, it's time to update the controller to first take a custom greeting as a parameter, and respond with the HTML template in the greet.html.eex file with the given greeting:

lib/example_server_html_eex/controller.ex

```
defmodule ExampleServerHtmlEex.Controller do
  # ..

  def greet(conn, %{"greeting" => greeting}) do
    conn
    |> put_status(200)
    |> render_html("greet.html.eex", greeting: greeting)
  end

  require EEx

  # wrapper around `Goldcrest.Controller.render/3` to
  # render html.eex templates from `priv/templates` folder.
  defp render_html(conn, file, assigns) do
    contents =
      file
      |> html_file_path()
      |> EEx.eval_file(assigns: assigns)

    render(conn, :html, contents)
  end

  # fetches html file from `priv/templates` directory
  defp html_file_path(file) do
    Path.join([
      :code.priv_dir(:example_server_html_eex),
      "templates",
      file
    ])
  end
end
```

In the preceding code snippet, we updated the ExampleServerHtmlEex.Controller. greet/2 function to match on the parameters. It now expects a greeting to be present when the greet action is invoked.

We have also defined two private functions, render_html/3 and html_file_path/1, that are responsible for fetching, reading, and evaluating an HTML template from the priv/templates folder and rendering its HTML contents, using EEx.SmartEngine's assigns.

```
$ iex -S mix
iex> opts = [
...>    scheme: :http,
...>    plug: ExampleServerHtmlEex.Router,
...>    options: [port: 4040]
...> ]
iex> %{start: {mod, fun, args}} = Plug.Cowboy.child_spec(opts)
iex> apply(mod, fun, args)
{:ok, #PID<0.313.0>}
```

Now, in a new browser window, let's visit the URL for greet with a greeting parameter, http://localhost:4040/greet?greeting=Hola. We should see something that looks like the following page:

Hola World!

Figure 6.2: The greet response with a custom greeting

Great! We have successfully managed to respond with a dynamic server-rendered HTML document using our web framework. Now, it's time to test out our controllers with the HTML response.

Testing our templates

In order to test our controllers with HTML responses, we will again leverage the Plug.Test module. Let's start by creating a test module:

---

test/example_server_html_eex/controller_test.exs

```
defmodule ExampleServerHtmlEex.ControllerTest do
  use ExUnit.Case
  use Plug.Test

  describe "GET /greet" do
    test "responds with an HTML document" do
      conn = conn(:get, "/greet?greeting=Hola")

      conn = ExampleServerHtmlEex.Router.call(conn, [])

      assert conn.status == 200
```

```
        assert conn.resp_body =~ "<h1>Hola World!</h1>"
      end
    end
end
```

In the preceding code, we make a GET request to /greet, with the greeting query parameter set to Hola. Therefore, we expect the resulting HTML to contain <h1>Hola World!</h1> and respond with a status of 200.

We can write similar tests for another use case where we don't send a greeting as a parameter:

test/example_server_html_eex/controller_test.exs

```
defmodule ExampleServerHtmlEex.ControllerTest do
    # ..

    test "raises error if no greeting provided" do
      conn = conn(:get, "/greet")

      assert_raise(Plug.Conn.WrapperError, fn ->
        ExampleServerHtmlEex.Router.call(conn, [])
      end)
    end
  end
end
```

When running the preceding test file, we should get two green tests:

$ mix test

..

Finished in 0.1 seconds (0.00s async, 0.1s sync)
2 tests, 0 failures

Randomized with seed 849460

In the preceding test file, we test the HTML body of the response as a string and check whether the string contains <h1>Hola World!</h1>. There are better ways to test HTML content using HTML parsers. Let's take a look at that in the next section.

Testing with Floki

In this section, we will update our controller test written in the previous section to test HTML elements returned by the response body. This will allow us to better ensure that the final HTML rendered by the controller is **correct**. In order to do this, we will use floki.

Floki is an HTML parser written purely in Elixir. With floki, we can parse an HTML string into a struct and search for notes using HTML and CSS selectors. It's a great tool to programmatically interpret HTML in Elixir, such as web scraping. One of its other common uses is to test server-side rendered HTML in Phoenix and other Elixir web applications.

Let's start by adding floki to the mix.exs dependencies:

mix.exs

```elixir
defmodule ExampleServerHtmlEex.MixProject do
  # ..

  defp deps do
    [
      {:goldcrest, "~> 0.0.1"},
      {:floki, ">= 0.0.0", only: :test}
    ]
  end
end
```

After running $ mix deps.get, we can update our controller test to test the actual content of the HTML returned by the GET /greet endpoint:

test/example_server_html_eex/controller_test.exs

```elixir
defmodule ExampleServerHtmlEex.ControllerTest do
  # ..

  describe "GET /greet" do
    test "responds with an HTML document" do
      conn = conn(:get, "/greet?greeting=Hola")

      conn = ExampleServerHtmlEex.Router.call(conn, [])

      assert conn.status == 200
      {:ok, html} = Floki.parse_document(conn.resp_body)

      heading =
        html
        |> Floki.find("h1")
        |> Floki.text()

      assert heading == "Hola World!"
    end
```

```
      end
    end
```

In the preceding test, we first called `Floki.parse_document/1` on the HTML response to ensure that the HTML document returned by `GET /greet` is valid. We then found the first element of type `h1` by using `Floki.find/2`. Finally, we piped the element into `Floki.text/1` to get the value of the heading. This is just one example of how Floki can help us write better tests for our HTML responses.

Summary

In this chapter, we first learned what `EEx` is. We covered several ways of using it, from dynamically evaluating a string to using a template file for a module's compilation. We also learned about `EEx.SmartEngine`, the default engine used by `EEx` for templating. We wrote a custom engine, `CustomEngine`, which used a marker that inspected the result of an element.

We then used all those concepts to update our HTTP server to return server-rendered HTML. We wrapped up by writing tests for our HTTP server and added `Floki`, an HTML parser, to write tests that allow us to validate an HTML response and test the attributes of an element.

In the next chapter, we will add the logic to define view modules, which will make using templates easier by housing some shared helper functions made accessible in the templates.

7
Working with Views

In the last few chapters, we learned how to render an HTML response for our web framework using the controller itself. In this chapter, we will learn about the view component that will be responsible for performing that rendering in a much better way. We will first extract some of the rendering and EEx logic to the new view module, and then define helper functions that will help us to render HTML and can be called in from the EEx template itself. We will then write a complete example application using the current state of our web framework.

The following are the topics that this chapter will cover:

- What is a view and what are its responsibilities?
- Defining a view module
- Calling a view module from the controller module
- Using EEx to pass helper functions at the time of evaluation
- Testing the view module
- Creating a new web application using `Goldcrest`

By the end, we will have an interface to simplify defining EEx templates with an appropriate place to house all the view helper functions.

Technical requirements

This chapter also relies on Elixir `1.12.x` and Erlang `23.2.x`. Most of this chapter includes extracting logic from the controller to the view component and exposing it in a way that makes sense. Therefore, while still recommended, it's not required you code along while reading this chapter.

In this chapter, I will also assume that you've read through and followed along with the previous chapter's code snippets. Since we will be expanding on how EEx engines work, it is important that you understand how EEx works and how we used EEx in the previous chapter to respond with server-rendered HTML.

Also, since we will be building our own views, any experience with views in a **Model-View-Controller (MVC)** framework such as Phoenix or Rails will be helpful, although not required.

The code examples for this chapter can be found at `https://github.com/PacktPublishing/Build-Your-Own-Web-Framework-in-Elixir/tree/main/chapter_07`

What is a view?

A view is the component of a web framework that is responsible for presenting data to an end user in a digestible way. Typically, in HTML-based web applications, a view is also responsible for handling the end user's interaction with the application.

In an MVC framework, a view is generally called from a controller implicitly, as with object-oriented web frameworks such as Rails and Django, or explicitly, as with functional web frameworks such as Phoenix. In *Chapter 4*, we explained how a request-response cycle works in a typical MVC flow. We can see how a view is called toward the end of the request phase and how it is responsible for presenting data in the response phase.

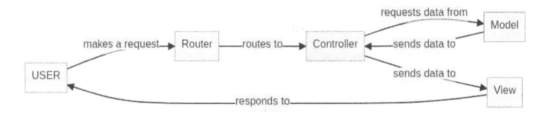

Figure 7.1: The MVC flow

In many web frameworks, views also have accompanying helper components such as templates, partials, or view components. Next, let's look at how views behave in Phoenix.

Views in Phoenix

In Phoenix, views act as helper modules, which are called from the controllers toward the end of a request and are responsible for rendering HTML using a template. In Phoenix, the views and templates have to comply with a strong naming convention. For example, `PageController` calls a `PageView` module to render templates present in the `web/templates/page/` folder by default.

By default, Phoenix generates `LayoutView` along with the `web/templates/layout` folder, containing a `root.html.eex` file (`root.html.heex` in newer versions). This template is rendered after every request, and within this template, other individual templates are rendered using other views. Let's take a look at an example of a Phoenix view and a template, for the `Page` resource.

Let's first look at the controller again:

```
defmodule ExampleWeb.PageController do
  use ExampleWeb, :controller

  def index(conn, params) do
    case Example.Page.load_data(params) do
      {:ok, data} ->
        render(conn, "index.html", data: data)

      {:error, error} ->
        conn
        |> put_flash(:error, error)
        |> redirect(to: "/")
    end
  end
end
```

The preceding controller calls the `render/3` function, which gets delegated to `PageView`, defined in the following code snippet, which further renders the `index.html.eex` template.

The following is an example of a view module:

```
defmodule ExampleWeb.PageView do
  use ExampleWeb, :view

  def heading, do: "Index Page"
end
```

In the aforementioned code snippet, we can see that a view module by default is just a simple module that inherits most of its behavior from the `use` macro. Doing this enables `PageView` to act like a Phoenix view and render templates with given assigns. We can also see that it defines a `heading/0` function, which simply responds with a string. To understand views better, let's look at the corresponding `index.html.eex`:

```
<h1><%= heading() %></h1>
```

We can see how the preceding HTML embeds Elixir, very similar to how we used EEx in the previous chapter. One thing worth noticing is how the `heading/0` function defined in the view module can be called in the `eex` template. This isn't something we have implemented yet, but it is a key feature in cleaning up our templates with reusable helper functions. We will implement this in the following sections.

Building a view interface

In order to build a view interface, let's first start with the example_server_html_eex app that we built in the previous chapter. In this app, we made the controller module responsible for both capturing a request from a router and rendering the final HTML. In this section, we will start by moving the rendering responsibility to Goldcrest.View.

We can start by defining the Goldcrest.View module that can house the render/3 function, which evaluates the EEx template and calls the render function in the Goldcrest.Controller module.

Let's create a new file in the goldcrest package:

lib/goldcrest/view.ex

```
defmodule Goldcrest.View do
  require EEx
  alias Goldcrest.Controller

  def render(conn, file, assigns) do
    contents =
      file
      |> html_file_path()
      |> EEx.eval_file(assigns: assigns)

    Controller.render(conn, :html, contents)
  end

  defp html_file_path(file) do
    # TODO
  end
end
```

In the preceding code snippet, we defined the Goldcrest.View module, which uses EEx.eval_file/2 to first evaluate the contents of the given file and then calls Goldcrest.Controller.render/3 to render the HTML contents. We also defined a private placeholder function, html_file_path, which is responsible for converting the given filename to a file path.

In order to define html_file_path, we need to first come up with a way to set a path to place all the templates for the application. In the case of example_server_html_eex, we placed the templates in the priv folder, but that might not be the case for all the applications. So, let's use the application-level configuration for goldcrest to define templates_path. With this approach, we can set templates_path for each application at the config level and use Application.fetch_env!/2 to get the configuration.

Let's implement html_file_path now:

lib/goldcrest/view.ex

```
defmodule Goldcrest.View do
  # ..

  defp html_file_path(file) do
    templates_path()
    |> Path.join(file)
  end

  defp templates_path do
    Application.fetch_env!(:goldcrest, :templates_path)
  end
end
```

Now that we have successfully moved the rendering responsibility to the view, we can update the
controller module in example_server_html_eex to just call Goldcrest.View directly:

lib/example_server_html_eex/controller.ex

```
defmodule ExampleServerHtmlEex.Controller do
  # ..

  # Update render to call `Goldcrest.View` here
  defp render(conn, file, assigns) do
    Goldcrest.View.render(conn, file, assigns)
  end
end
```

In Phoenix, we saw how each resource had a controller and a view module, so in order to keep that
pattern, let's move the responsibility of rendering to a view module:

lib/example_server_html_eex/controller.ex

```
defmodule ExampleServerHtmlEex.Controller do

  # Update render to call `ExampleServerHtmlEex.View` here
  defp render(conn, file, assigns) do
    ExampleServerHtmlEex.View.render(conn, file, assigns)
  end
end
```

Now, let's define our view module with a `render/3` function that simply delegates to `Goldcrest. View.render/3`:

lib/example_server_html_eex/view.ex

```
defmodule ExampleServerHtmlEex.View do
  def render(conn, file, assigns) do
    Goldcrest.View.render(conn, file, assigns)
  end
end
```

Now, we have successfully moved our functions to define a central view interface for the controller. Next, we will update `Goldcrest.View.render/3` to allow us to pass helper functions from our view module to the templates, similar to Phoenix.

Passing helper functions to templates

We saw earlier in the chapter how Phoenix allows us to use functions defined in the view module inside the templates. This further allows us to clean up the templates and extract shared/complex logic in the templates, by moving it to the views. In this section, we will learn how we can use the EEx module to pass functions to the HTML templates.

In the previous chapter, we used the first two arguments of `EEx.eval_file/3` to pass a file path and list of variable bindings to the template. The third argument of this function is a list of options, one of which is `:functions`. By using the `:functions` key, we can pass a list of functions that are accessible to the template at the time of its evaluation. We can use this feature to pass a list of functions defined in the view module to our HTML template.

Now that we know how to pass a list of functions to a template, we just need a way to get a list of all the public functions defined in the view module. Once we have that list, we can pass the list of functions as a value to the `:functions` option for the `EEx.eval_file/3` call. We can use the `Module.__info__/1` function to get that information.

The__info__1 callback

Elixir defines a function __info__1 on every module that's defined in memory. The purpose of this function is to get information related to the metadata of every module. This includes functions, macros, attributes, hashes , and so on.

The __info__1 function takes one atom as the argument. The atom can be one of the following:

:attributes: To get a list of persisted module attributes defined

:functions: To get a list of public functions along with their arity

:macros: To get a list of public macros along with their arity

:module: To get the atomified name of a module

:md5: To get the unique MD5 hash of the module

:compile: To get the module's compile-time metadata

Now, let's update `Goldcrest.View` to take the view module as the first argument. This will allow us to pass a list of functions defined in the view module at the time of the template evaluation:

lib/goldcrest/view.ex

```
defmodule Goldcrest.View do
  # ..

  def render(view_module, conn, file, assigns) do
    functions = view_module.__info__(:functions)

    contents =
      file
      |> html_file_path()
      |> EEx.eval_file(
        [assigns: assigns],
        functions: [
          {view_module, functions}
        ]
      )

    Controller.render(conn, :html, contents)
  end
end
```

In the preceding code snippet, we used the `__info__/1` reflection function for the given view module to evaluate a template using EEx, which has access to all the public functions defined in the view module. Now, we can update the `render` function in the `ExampleServerHtmlEex.View` module to send itself as the first argument to the `Goldcrest.View.render/4` function:

lib/example_server_html_eex/view.ex

```
defmodule ExampleServerHtmlEex.View do
  def render(conn, file, assigns) do
    # sends `__MODULE__` as the first argument.
    Goldcrest.View.render(__MODULE__, conn, file, assigns)
  end
end
```

Now, we can use our view module as a helper module whose functions can be accessed in the templates for better organization. I know it's not ideal to have to manually pass the module name to the render function, but we can improve the interface in *Part 3* of the book, where we will learn about metaprogramming and build a cleaner interface around all our components.

Taking Goldcrest for a spin

In this section, we will use what we've built, and the features we've added to Goldcrest up to this point, to build an example web application. This application will support routes that respond with both HTML and JSON. We will also test out the new feature added to `Goldcrest.View`, which allows us to call a helper function inside an HTML template.

Let's start by creating a new mix project and adding `goldcrest` to its `mix` dependencies:

mix.exs

```
$ mix new tasks_web --sup
..
$ cd tasks_web
defmodule TasksWeb.MixProject do
  # ..

  # Run "mix help deps" to learn about dependencies.
  defp deps do
    [
      {:goldcrest, path: "../goldcrest"}
    ]
  end
end
```

Running `mix deps.get` should resolve `hex` dependencies for the `tasks_web` app.

Now, let's add a router module with a default 404 handler. This should be very similar to the one we added in earlier chapters:

lib/tasks_web/router.ex

```
defmodule TasksWeb.Router do
  use Plug.Router

  plug Plug.Parsers,
    parsers: [:urlencoded, :multipart],
    pass: ["text/html", "application/*"]

  plug :match
  plug :dispatch

  match _ do
    send_resp(conn, 404, "<h1>Not Found</h1>")
  end
end
```

In order for our requests to reach our router, we need to hook it up to a web server. Let's update our application supervision tree to start an HTTP server that routes to `TasksWeb.Router`:

lib/tasks_web/application.ex

```
defmodule TasksWeb.Application do
  # ..

  @impl true
  def start(_type, _args) do
    children = [
      {
        Plug.Goldcrest.HTTPServer,
        [
          plug: TasksWeb.Router,
          port: 4040,
          options: []
        ]
      }
    ]

    # See https://hexdocs.pm/elixir/Supervisor.html
    # for other strategies and supported options
```

```
    opts = [strategy: :one_for_one,
            name: TasksWeb.Supervisor]
    Supervisor.start_link(children, opts)
  end
end
```

Now, we have a web server running, which simply responds with a 404 message to all incoming requests. Let's start our server and try it:

```
$ mix run --no-halt
01:02:40.650 [info]  Started a webserver on port 4040
```

In a separate terminal window, we can make a request to our running server:

```
$ curl http://localhost:4040
<h1>Not Found</h1>
```

Let's now add a new controller, TasksWeb.TaskController, which is a way to manage a task resource. In order to keep our infrastructure simple, we will avoid the use of any database or ecto; therefore, we will store tasks in a running process using the Agent interface.

We will define three actions in our TasksWeb.TaskController:

- GET index (HTML): This will be an HTML interface to get a list of all the tasks added to our application.

- POST create (HTML): This will be an HTML interface that provides a way to add tasks to our application.

- GET delete (JSON): This will also be an HTML interface, to remove tasks from our application. We're making this action GET instead of DELETE to ensure easier HTML in the latter part of the chapter.

In order to support the preceding actions, we will first have to add the corresponding routes to our router:

lib/tasks_web/router.ex

```
defmodule TasksWeb.Router do
  # ..
  alias TasksWeb.TaskController

  get "/tasks/:id/delete", do: TaskController.call(conn,
    action: :delete)
  get "/tasks", do: TaskController.call(conn,
    action: :index)
  post "/tasks", do: TaskController.call(conn,
    action: :create)
```

```
  match _ do
    send_resp(conn, 404, "<h1>Not Found</h1>")
  end
end
```

In the preceding code snippet, we add three new routes to the Router module, which delegate to the appropriate TaskController's actions. As usual, we also add a general match clause for any requests that don't match any defined routes. This simply responds with a 404 HTML page.

Defining a store

In order to properly implement these controller actions, let's first define our store. As mentioned earlier, we will use Elixir's Agent interface to store a list of tasks, represented by a two-element tuple (a name and a description). The best way to store them is in a long-running process, which is part of the application's supervision tree. Let's define a TasksWeb.Tasks module, which provides an interface to interact with the Agent process that will store all the tasks.

We will use three functions defined in the Agent module:

- Agent.start_link/2: This is used to start a long-running process with an initial state and given name. We will be using the name of the module (TasksWeb.Tasks) for easier access in future functions.

- Agent.get/2: This is used to list all the elements (tasks) stored in the agent with a given name.

- Agent.update/2: This is used to add or delete an element from the list of elements stored in the agent with the given name:

lib/tasks_web/tasks.ex

```
defmodule TasksWeb.Tasks do
  @moduledoc """
  In-memory, volatile store for tasks. Relies on Agent
  interface.
  """

  def start_link, do: Agent.start_link(fn -> [] end,
    name: __MODULE__)

  def child_spec(opts) do
    %{
      id: __MODULE__,
      start: {__MODULE__, :start_link, opts},
      type: :worker,
```

```
        restart: :permanent,
        shutdown: 500
      }
    end

    def list, do: Agent.get(__MODULE__, & &1)

    def add(name, description) do
      Agent.update(__MODULE__, & &1 ++ [{name, description}])
    end

    def delete(index) do
      Agent.update(__MODULE__, &List.delete_at(&1, index))
    end
  end
```

In the preceding code snippet, we define an interface to manage the process, which stores a list of tasks as its state. We define functions such as start_link/0, add/2, delete/1, and list/0 to interact with the process and perform **Create Read Update Delete (CRUD)** actions on the list of tasks. Furthermore, we define a child_spec/1 function, which returns a specification that can be used to add this process to Supervisor as part of the application's supervision tree.

Next, let's use the child_spec/1 function to add TasksWeb.Tasks to the supervision tree children defined in the TasksWeb.Application module:

lib/tasks_web/application.ex

```
defmodule TasksWeb.Application do
  # ..

  @impl true
  def start(_type, _args) do
    children = [
      {
        Plug.Goldcrest.HTTPServer,
        [
          plug: TasksWeb.Router,
          port: 4040,
          options: []
        ]
      },
      {
        TasksWeb.Tasks,
        []
      }
```

```
    ]

    # See https://hexdocs.pm/elixir/Supervisor.html
    # for other strategies and supported options
    opts = [strategy: :one_for_one,
            name: TasksWeb.Supervisor]
    Supervisor.start_link(children, opts)
  end
end
```

Similar to `Plug.Goldcrest.HTTPServer`, we've also added `TasksWeb.Tasks` to the list of children being supervised by `TasksWeb.Supervisor`. In the case of `TasksWeb.Tasks`, we've provided an empty list as options, which get delegated to the initial list of tasks when the process is started. This will ensure that on the application's startup, the state of the `TasksWeb.Tasks` process is an empty list.

Now that we have a way to store and interact with the tasks stored in our application, we can call the functions defined in the `TasksWeb.Tasks` module in the `TasksWeb.TaskController` module as part of the controller actions.

The index action

Let's start by defining the `index` action. We can take most of the code in this new module from the code provided in the `ExampleServerHtmlEex.Controller` module. This includes the `use` and `import` statements, the `call/2` and `render/3` functions, and the arity of the controller actions. Once we have those functions, we can define the `index/2` function in the `TasksWeb.TaskController` module:

lib/tasks_web/controllers/task_controller.ex

```
defmodule TasksWeb.TaskController do
  use Plug.Builder
  import Plug.Conn
  alias TasksWeb.Tasks

  # Taken from other examples
  def call(conn, action: action) do
    conn = super(conn, [])

    apply(__MODULE__, action, [conn, conn.params])
  end

  def index(conn, _params) do
    tasks = Tasks.list()
```

```
    conn
    |> put_status(200)
    |> render("tasks.html.eex", tasks: tasks)
  end

  # Taken from other examples
  defp render(conn, file, assigns) do
    TasksWeb.TaskView.render(conn, file, assigns)
  end
end
```

In the preceding code snippet, we've taken some code from other example controllers as indicated previously and have defined an `index/2` function, which corresponds to the `get` action in the controller. In the action, we use the `TasksWeb.Tasks` module's function, `list/0`, to get the list of tasks defined in the module, and we call `render/2` with the name of the `html.eex` file to be rendered.

In order for our controller to now compile, we must define a view module, `TasksWeb.TaskView`, which defines the `render/3` function, much like the `ExampleServerHtmlEex.View` module:

lib/tasks_web/view/task_view.ex

```
defmodule TasksWeb.TaskView do
  def render(conn, file, assigns) do
    Goldcrest.View.render(__MODULE__, conn, file, assigns)
  end
end
```

The preceding code snippet is identical to the `ExampleServerHtmlEex.View` module, as it simply delegates the `render/3` function to the `Goldcrest.View.render/4` function.

Next, we need to update the configurations to specify `templates_path` as part of our mix project, `:goldcrest`:

config/config.exs

```
use Mix.Config

config :goldcrest,
  templates_path: "priv/templates"
```

Now that we have set `templates_path` to `"priv/templates"`, we can add the template that is rendered as part of the `index` action of the controller, `tasks.html.eex`, in the `priv/templates` folder:

priv/templates/tasks.html.eex

```
<table>
  <thead>
    <tr>
      <th>Name</th>
      <th>Description</th>
    </tr>
  </thead>
  <tbody>
  <%= for {name, description} <- @tasks do %>
    <tr>
      <td><%= name %></td>
      <td><%= description %></td>
    </tr>
  <% end %>
  </tbody>
</table>
```

In the preceding code snippet, we simply use a list comprehension (`for`) to render a list of tasks in a list of table rows. This file looks very similar to a `Phoenix` template, where the list of tasks is accessed by the `@tasks` attribute, which is passed to the template at the time of its evaluation by `EEx.SmartEngine`, which was covered in the previous chapter.

Now, let's test out our web application. We can start our application by running `mix run --no-halt`.

Let's first hit a route that doesn't exist, `/` (`root`). We should see a `404` response with the HTML body saying **Not Found**.

Figure 7.2: The Not Found request

Now, let's try making a request to the GET /tasks route. This should respond with an HTML table with zero rows, since we haven't added any tasks to our app yet.

Name Description

Figure 7.3: The empty tasks request

The create action

Now that we have a way to list all the tasks in our store, let's come up with an interface to add a new task. We can start by adding a form as part of the HTML rendered by the index action itself. Submitting this form should invoke a new controller action, create.

So, let's update our tasks.html.eex template to render a form:

priv/templates/tasks.html.eex

```
<h1>Listing Tasks</h1>
<table>
  <thead>
    <tr>
      <th>Name</th>
      <th>Description</th>
    </tr>
  </thead>
  <tbody>
  <%= for {name, description} <- @tasks do %>
    <tr>
      <td><%= name %></td>
      <td><%= description %></td>
    </tr>
  <% end %>
  </tbody>
</table>

<br />

<hr />

<h1>Add new Task</h1>
<form method="POST" action="/tasks">
```

```
<label for="name">Name:</label>
<br />
<input type="text" id="name" name="name">
<br><br>

<label for="description">Description:</label>
<br>
<textarea id="description" name="description">
</textarea>
<br><br>

<input type="submit">
</form>
```

In the preceding code snippet, we add a new section to the HTML, rendered as part of the GET /
tasks request. This new section is a form that POST to the /tasks path with the form parameters.

Let's try making the same GET /tasks request again by restarting our web application.

```
$ mix run --no-halt
```

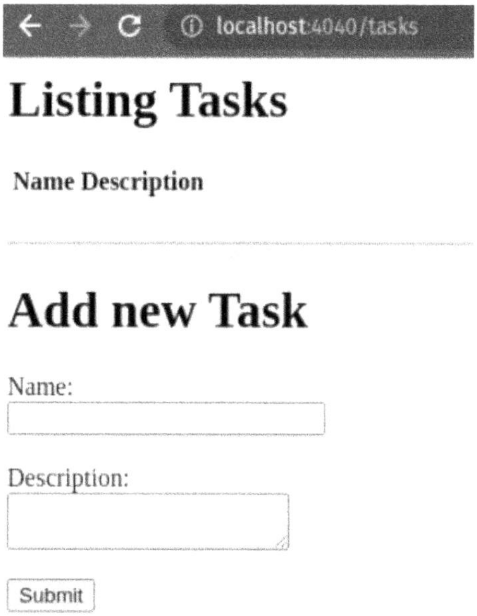

Figure 7.4: Tasks with a form

Now that we have a form, let's define a controller action, `create/2`, which takes parameters from this form and inserts a new task in our store:

lib/tasks_web/controllers/task_controller.ex

```elixir
defmodule TasksWeb.TaskController do
  use Plug.Builder
  import Plug.Conn
  alias TasksWeb.Tasks
  def create(conn, %{"name" => name, "description" =>
    description}) do
    Tasks.add(name, description)

    conn
    |> Goldcrest.Controller.redirect(to: "/tasks")
  end
end
```

In the preceding code snippet, we define a `create/2` action that simply takes two parameters – a name and a description. It uses those two parameters coming from the form to create a new task and then redirects back to the `index` action.

It's time to restart our app and try submitting a form to create a new task. Since the action redirects back to the index page, we should see a new task that is created by submitting the form.

```
$ mix run --no-halt
```

Figure 7.5: Task form submission

Now, let's submit the preceding form!

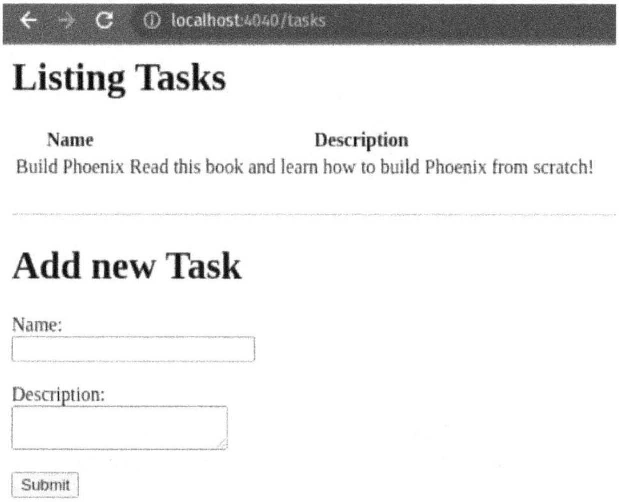

Figure 7.6: Task form post-submission

In the preceding screenshot, we can see that after submitting the newly created form, a new row is created in the tasks table, indicating that the task was added to our store.

The delete action

Now that we have a way to create and list all the tasks, all that's left in our routes is to handle the deletion of a task.

Similar to the `create` action, let's start by adding the HTML interface to delete a task. In this case, a link next to each row in the index table would be the easiest way to implement this.

Let's update `tasks.html.eex` again:

priv/templates/tasks.html.eex

```
<h1>Listing Tasks</h1>
<table>
  <thead>
    <tr>
      <th>Name</th>
      <th>Description</th>
      <th></th>
    </tr>
  </thead>
```

```
  <tbody>
  <%= for {{name, description}, index} <-
    Enum.with_index(@tasks) do %>
    <tr>
      <td><%= name %></td>
      <td><%= description %></td>
      <td>
        <a href="/tasks/<%= index %>/delete">Delete</a>
      </td>
    </tr>
  <% end %>
  </tbody>
</table>
</form>
```

In the preceding code snippet, we add a new header with no text and a new column to reach the row of the tasks table. This column will contain a link to delete a specific task using its index.

Let's render this page with a new task. We should see a similar page to the one shown here.

Figure 7.7: Task with a Delete link

Now, let's define the `delete` action in the `TaskController` module:

lib/tasks_web/controllers/task_controller.ex

```
defmodule TasksWeb.TaskController do
  use Plug.Builder
  import Plug.Conn
  alias TasksWeb.Tasks
  def delete(conn, %{"id" => id}) do
    id
    |> String.to_integer()
    |> Tasks.delete()

    conn
    |> Goldcrest.Controller.redirect(to: "/tasks")
  end
end
```

Now, let's restart our app again, add a new task, and try clicking the **Delete** link:

```
$ mix run --no-halt
```

Figure 7.8: Empty task

A new task is added with the **Delete** link.

Figure 7.9: After clicking the Delete link

Now that we have implemented all three controller actions, this concludes the features we will be adding to our web framework. From here onwards, we will leverage one of Elixir's most powerful features, metaprogramming, to wrap everything we've built within a simple, idiomatic interface, just like Phoenix.

Summary

In this chapter, we learned the concept of a view and a view helper. We learned how Phoenix implements a view and how it makes defining templates a lot easier. We then learned how we can pass helper functions to an EEx template and used that information to define our own view interface.

Finally, we wrapped up by combining all the previously written code and building a fully functional web application with an in-memory data store, using Elixir's Agent interface. This chapter also marks an end to the non-metaprogramming parts of this book. From here on out, our goal is to define a better and reusable interface for all the code we've written until now.

Exercises

While building the TasksWeb application, we didn't write tests for all the routes and requests. We instead relied on viewing the final rendered HTML. How would you go about testing all the routes and controller actions defined in this app? What use cases and edge cases would you like to test?

Part 3: DSL Design

In this part, you will learn about metaprogramming in Elixir and how to wrap the functionality built in *Part 2* in an interface that's easy to use and read.

This part includes the following chapters:

8

Metaprogramming – Code That Writes Code

In this chapter, we will learn about one of the most powerful features of Elixir, metaprogramming. Writing a simple piece of Elixir code is referred to as *programming*, and writing Elixir code that programmatically injects behavior into other Elixir code is referred to as *metaprogramming*. Metaprogramming allows us to extend Elixir's features and facilitate developer productivity by automating the otherwise tedious and manual process of adding repetitive boilerplate code.

Let's take the Phoenix router, for example:

---

blog/lib/blog_web/router.ex

```
defmodule BlogWeb.Router do
  # ..
  pipeline :browser do
    plug :fetch_session
    plug :accepts, ["html"]
  end

  scope "/", BlogWeb do
    pipe_through [:browser]

    get "/posts", PostController, :index
  end
end
```

The preceding **Domain - Specific Language (DSL)** is only possible due to metaprogramming. On the surface, this code looks easy to follow and understand, but there is a lot going on under the hood.

In this chapter, we will uncover some of the constructs that allow us to write DSLs, such as the Phoenix router. We will not only cover the fundamentals of metaprogramming in Elixir but also understand some of the risks involved in using it, and how to minimize them. We will then finish by covering different ways to test that code.

We will cover the following topics in this chapter:

- The pros and cons of metaprogramming
- When to use metaprogramming
- How to use metaprogramming sustainably
- Abstract syntax trees
- Quoted literals in Elixir
- Code injection
- The hygienic evaluation of quoted literals
- Macros
- The Elixir compilation process
- Compile-time callbacks
- Building a DSL in Elixir

Technical requirements

This chapter is filled with a lot of `iex` examples, along with the usual Elixir code snippets. To get the most out of this chapter, I recommend following along with all the `iex` examples and trying to change a few things to see how that changes the final results.

Just like the rest of this book, this code in this chapter was run using Elixir `1.12.x` and Erlang `23.2.x`.

This chapter is one of the harder chapters to understand in this book because of the nature of metaprogramming. Therefore, reading it multiple times and experimenting with the code in the chapter is the key to setting up a good metaprogramming foundation for the rest of the book.

The code examples for this chapter can be found at `https://github.com/PacktPublishing/Build-Your-Own-Web-Framework-in-Elixir/tree/main/chapter_08`

The pros and cons of metaprogramming

Let's take a look at the router example again:

---

blog/lib/blog_web/router.ex

```
defmodule BlogWeb.Router do
  # ..
  pipeline :browser do
    plug :fetch_session
    plug :accepts, ["html"]
  end

  scope "/", BlogWeb do
    pipe_through [:browser]

    get "/posts", PostController, :index
  end
end
```

In the preceding code example, we can easily define a /posts route that accepts html requests and routes requests to the PostController.index/2 action. The code is concise, standardized, and easy to digest.

Here are the pros of metaprogramming. Metaprogramming can do the following:

- Hide the complexity of implementation under a simple and concise layer of abstraction. This is often referred to as a DSL.

- Increase developer productivity because the amount of code that needs to be written could be drastically reduced, as seen in the preceding router example.

- Automate and standardize repetitive boilerplate code. This way, we won't have to worry about supporting multiple implementations of the same problem because it is enforced via a simpler interface.

Metaprogramming, as with any other powerful feature, has a lot of cons as well. The following are some of those:

- Metaprogramming decreases code transparency. This comes hand in hand with hiding the complexity of implementation. Since it's easier to digest the DSL part of code, it becomes a lot harder to understand the implementation. This also means it can be very hard to do things that don't follow a convention.

- Metaprogramming increases overall code complexity. This should be an expected result because when you write metaprogrammed code, you're writing code of a higher order. It's code that takes other code as arguments, and therefore, it is oftentimes harder to test, maintain, and debug.

Metaprogramming is extremely powerful, and with this great power comes great responsibility. This responsibility mostly entails using it thoughtfully.

When to use metaprogramming

Now that we know that metaprogramming has its own pros and cons, it is very important to learn about when to use it and how to decide whether you need it or not. The simplest answer to this question is almost never. In other words, before even considering metaprogramming, you should try to implement your solution without it. There are two advantages to this:

- You will most likely write a good, less complex solution that doesn't rely on metaprogramming, and has low maintenance costs.

- You will write an almost complete solution that will exist independent of any metaprogrammed code. This also sets you up to decouple your interface (the metaprogrammed code) from your implementation (the non-metaprogrammed code).

Here are some other points for you to consider when considering metaprogramming. Use metaprogramming in the following situations:

- When you have exhausted all the other options. As mentioned previously, metaprogramming is complex and should be used as the very last resort.

- When you are writing a DSL that will be used in multiple places. Using metaprogramming prematurely is one of the most common mistakes developers make. If you have requirements for DSL that will be used only for one project, there is generally no need to wrap it in an interface (especially a metaprogrammed interface). Waiting for future implementations allows you to first see whether you need to write a DSL to avoid copying the code, and also gauge how the requirements change between those use cases. This will further allow you to better define your interface. Remember that flexibility comes at the cost of complexity, so there's no need to write a flexible interface prematurely.

- When you have minimized metaprogrammed code – that is, separated your interface from your implementation. This will significantly cut the cost of maintaining the code base.

- When you have maximized its determinism by adding thorough unit tests and integration tests. Since metaprogramming is more complex than your everyday code, it relies a lot more on automated tests. If your interface is complex enough, it might even make sense to add property-based tests to increase your confidence in its behavior. This is also another place where separating the interface from the implementation will help. This way, you can more easily test your metaprogrammed code (the interface) without necessarily having to test your implementation.

- When you have maximized its inspectability. Most metaprograms are not inspectable. Therefore, they are not very debuggable. Putting thought into adding introspective features to a DSL could significantly decrease the pain of debugging and extending the interface.

- When the cost of failure is minimal. It's always easier to use metaprogramming for something such as tests than production code. This is not to say that tests are not important, but at least the cost of failure for them wouldn't suddenly impact your application.

To summarize, metaprogramming is a great tool and it's not **evil**. However, it needs to be used very thoughtfully, with an understanding that you're adding code of a higher order of complexity.

Metaprogramming in Elixir

In Elixir, metaprogramming is mostly used as a way to extend the language's functionality and design DSL to make code more readable and digestible. However, unlike many other metaprogrammable languages, Elixir metaprogramming is quite restrictive. There are some safety mechanisms baked into Elixir's core library to prevent developers from doing something that could easily break the language. For example, you cannot define a function in a module after it has already been defined. We will cover more such mechanisms later in this chapter.

Metaprogramming in Elixir revolves around three main pillars:

- **Quoted literals**: The Elixir representation of an **Abstract Syntax Tree** (**AST**) of an Elixir program

- **Dynamic code injection**: Injecting behavior into existing code dynamically

- **Compile-time callbacks**: Updating a module or the behavior of a function before or after its elements are defined

We will first start by understanding how an AST is represented in Elixir.

Abstract Syntax Tree

An Abstract Syntax Tree is a tree representation of the structure of source code written in a programming language. This tree doesn't represent all of the detail that appears in the code, but just shows the basic structure that would impact its execution.

The tree is generated using the process of lexical analysis of code, also known as tokenization, followed by the process of syntactical analysis, also known as parsing. Tokenization gets a list of tokens (functions, operators, etc.) and, as part of the parsing process, those tokens are added as nodes to a tree, which is the Abstract Syntax Tree.

For example, consider the following pseudo code:

```
log(pow(4, 2) + 5, 3)
```

The preceding code will generate the following AST:

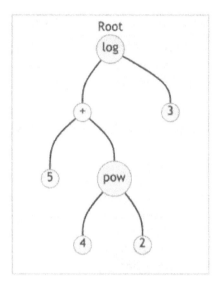

Figure 8.1: The AST for log(pow(4, 2) + 5, 3)

In the preceding diagram, we can see the tree representation of tokens generated after the lexical analysis of the log (pow (4, 2) + 5, 3) expression, where each node is a function or a nested function, and each leaf is an argument to the function.

Now that we understand what an AST is, let's take a look at quoted literals, which are an AST representation of Elixir code.

Understanding quoted literals

To properly understand metaprogramming in Elixir, you first have to see how Elixir represents its code internally, and to do that, you have to understand the structure of an AST in Elixir. For any Elixir code, there exists an AST representation in Elixir. This feature helps Elixir to be bootstrapped, where most of Elixir is written in Elixir itself. Note that the Elixir representation of an AST is different from the final AST generated by the Elixir code.

The Elixir representation of a program's abstract syntax tree is referred to as *quoted literals* or *quoted expressions*. Internally, every quoted literal in Elixir is composed of a three-element tuple.

Let's look at the following code, for example:

```
{:*, [context: Elixir, import: Kernel], [a, b]} # AST for a * b
```

In the preceding code, we use a three-element tuple to represent an AST:

- The first element is the function being called. In this case, it is `:*`.

- The second element is the metadata. This has information about what modules are imported and in what context an expression was quoted.

- The third element is the list of arguments for the function in the first element of the tuple. In this case, it is a and b.

Elixir also provides the `quote` construct to convert vanilla Elixir code to its AST representation. Let's take a look at the following code:

```
iex> quote do: 1 * 2
{:*, [context: Elixir, import: Kernel], [1, 2]}
```

The preceding snippet shows the quoted version of an Elixir expression obtained using the `quote/2` macro. It can be represented as a tree, using the first and the last elements of the tuple, as follows:

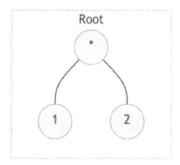

Figure 8.2: AST for 1 * 2

Now that we understand how an AST is represented in Elixir, we can take a look at a quoted literal for a more complex block of code:

```
iex shell
iex> quote do
...>    import IO
...>    def hello, do: puts "Hello World!"
...> end
{:__block__, [],
  [
    {:import, [context: Elixir], [{:__aliases__,
      [alias: false], [:IO]}]},
    {:def, [context: Elixir, import: Kernel],
      [{:hello, [context: Elixir], Elixir}, [do: {:puts, [],
```

```
        ["Hello World!"]}]]}
  ]}
```

Let's represent the final quoted literal in an AST, using the first and third elements of all the tuples.

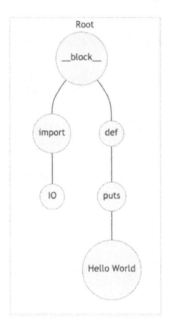

Figure 8.3: AST for the above quoted literal

As you can see in the preceding figure, the quoted literal follows a tree-like structure where the first element is the token (an operator or a function), the second element is metadata related to the code, and the third element is a list of arguments that can also be another quoted literal as a three-element tuple.

Evaluating quoted literals

In Elixir, we can quickly evaluate a quoted literal using the Code.eval_quoted/3 function. So, let's use that function to better understand quoted literals:

```
iex shell
iex> quoted_expr = quote do: 1 + 1
{:+, [context: Elixir, import: Kernel], [1, 1]}
iex> Code.eval_quoted(quoted_expr)
{2, []}
```

The preceding quoted literal gets evaluated as expected. You might have noticed that `Code.eval_quoted/3` returns a two-element tuple. The first element is the result of the evaluation, and the second element is `binding` (variables defined) with which the quoted literal was evaluated. Now, let's introduce some dynamic behavior to the expression:

```
iex shell
iex> quoted_expr = quote do: 1 + b
{:+, [context: Elixir, import: Kernel], [1, {:b, [], Elixir}]}
iex> Code.eval_quoted(quoted_expr)
** (CompileError) nofile:1: undefined function b/0
```

The preceding code shows us a variable to the quoted literal. The variable needs to be defined during the time of its evaluation. A quick way of doing that is to merge the quoted literal with another one that defines the b variable, another quoted literal (a quoted Elixir block):

```
iex shell
iex> quoted_expr = quote do: 1 + b
{:+, [context: Elixir, import: Kernel], [1, {:b, [], Elixir}]}
iex> quoted_expr2 = quote do: b = 2
{:=, [], [{:b, [], Elixir}, 2]}
iex> container_quoted_expr = {:__block__, [], [quoted_expr2, quoted_
expr]}
{:__block__, [],
 [
   {:=, [], [{:b, [], Elixir}, 2]},
   {:+, [context: Elixir, import: Kernel], [1, {:b, [], Elixir}]}
 ]}
iex> Code.eval_quoted(container_quoted_expr)
{3, [{{:b, Elixir}, 2}]}
```

The preceding code returns what the expression evaluates to, 3, along with the variable bindings. Note how we added `quoted_expr2` to the container before `quoted_expr`. This is because we need the b variable defined before we call `1 + b`.

The hygienic evaluation of quotes

In Elixir, quoted literals are evaluated hygienically. This means variables defined outside the scope of the quoted literals do not get defined inside the quoted literal scope. It also means that the variables defined inside the quoted literals do not get defined outside the scope, even upon evaluation.

Let's take a look at the following code:

```
iex shell
iex> expr = quote do: b = 2
{:=, [], [{:b, [], Elixir}, 2]}
```

```
iex> Code.eval_quoted(expr)
{2, [{{:b, Elixir}, 2}]}
iex> b
** (CompileError) iex: undefined function b/0
```

As you can see in the preceding code, anything defined inside a quoted literal upon its evaluation isn't defined outside its scope. Even though the b variable was defined inside the scope of the expression when it was evaluated, trying to reference it outside that scope raised an error.

Now, let's look at the following code snippet:

```
iex shell
iex> a = 1
1
iex> expr = quote do: 1 + a
{:+, [context: Elixir, import: Kernel], [1, {:a, [], Elixir}]}
iex> Code.eval_quoted(expr)
warning: variable "a" does not exist and is being expanded to "a()",
please use
parentheses to remove the ambiguity or change the variable name
  nofile:1

** (CompileError) nofile:1: undefined function a/0
```

The preceding code snippet shows that a variable defined outside the scope of a quoted literal cannot be referenced inside the quoted literal. We can see that a was defined outside the quoted literal, but referencing a inside the quoted literal raised an error upon evaluating it.

The hygienic evaluation of quoted literals was added to ensure that variables don't leak in and out of the scopes. This is to ensure that quoted blocks behave the way you expect them to at the time of their definition and not depending on the environment in which they are evaluated.

Even though hygiene was added to quote evaluation as a safety mechanism, there are ways to bypass it to add more dynamic behavior to quoted literals.

Adding dynamic behavior to quotes

As you might guess, it's not feasible to merge two quoted literals manually every time we need to add dynamic behavior to one. So, Elixir provides other constructs that allow us to manipulate a quoted literal. We're going to closely look at two of those next – unquote and var!.

In Elixir, unquote is a way to add behavior to a quoted literal at the time of its definition. For example, the following code uses unquote inside a quoted literal:

```
iex shell
iex> b = 3
iex> quoted_expr = quote do: 1 + unquote(b)
{:+, [context: Elixir, import: Kernel], [1, 3]}
```

As you can see, instead of returning [1, b] as the third element of the tuple, unquote evaluates the value of b to return [1, 3] as the third element. Therefore, unquote updates a quoted literal at the time of its definition.

We can use var! to add behavior to a quoted literal at the time of its evaluation.

Now, consider the following code:

```
iex shell
iex> b = 3
iex> quoted_expr = quote do: 1 + var!(b)
{:+, [context: Elixir, import: Kernel],
 [1, {:var!, [context: Elixir, import: Kernel], [{:b, [], Elixir}]}]}
iex> Code.eval_quoted(quoted_expr, b: b)
{4, [b: 3]}
```

The preceding code uses var!/2, which allows us to evaluate a quoted literal at the time of its evaluation, using the binding passed to Code.eval_quoted/3.

Unhygienic evaluation with var!

var! is a way to make an evaluation of a quoted literal (or an expansion of a macro) unhygienic. This could lead to the quoted and unquoted environments conflicting with each other and result in unexpected behavior of the final code. Therefore, it is important to be very careful before using var!, as it could directly change the user environment.

Now that we know how to dynamically evaluate a quoted literal in a hygienic way, let's put it to the test by using it to inject behavior into a module at the time of its compilation.

Code injection using evaluation

Now that we understand how Code.eval_quoted/3 works, we can use that function to inject behavior into a module from another module.

The following is a module that defines a function that returns a quoted literal. The quoted literal, when evaluated, further defines a hello function inside the module it is evaluated in. The function is defined as a simple hello world function:

behavior_injector.ex

```
defmodule BehaviorInjector do
  def behaviour_quoted_expr do
    quote do
      def hello, do: IO.puts "Hello world!"
    end
  end
end
```

Now, let's use the previously defined module's function inside another module:

test_subject.ex

```
defmodule TestSubject do
  Code.eval_quoted(BehaviorInjector.behaviour_quoted_expr(), [], __ENV__)
end
```

In the preceding code snippet, we need to provide __ENV__ in order to pass compile-time information of the TestSubject module to properly inject behavior inside it. Now, it's time to try it out in iex:

```
iex shell
iex> c "./behavior_injector.ex"
[BehaviorInjector]
iex> c "./test_subject.ex"
[TestSubject]
iex> TestSubject.hello()
Hello world!
:ok
```

We were able to successfully inject behavior from one module to another using the preceding method. However, there is a cleaner (and correct) way to do it using the macro construct.

Code injection using macros

We learned that we can use `Code.eval_quoted/3` to dynamically add behavior to a module, but Elixir provides a cleaner and more consistent way of doing this, using `macro`.

`Code.eval_quoted/3` allows us to inject code using a set of variable bindings and an environment. `macro`, on the other hand, allows us to define a set of quoted literals at the time of their compilation and evaluate the expressions, by simply invoking it like a function inside another module. A macro can be defined by using `defmacro/2`, which itself is a macro.

To understand the preceding distinction better, let us use `macro` to implement the same `BehaviorInjector` module and inject behavior in the `TestSubject` module:

behavior_injector.ex

```
defmodule BehaviorInjector do
  defmacro define_hello do
    quote do
      def hello, do: IO.puts "Hello world!"
    end
  end
end
```

In the preceding code snippet, we simply replaced a function thats returns a quoted expression with a macro:

test_subject.ex

```
defmodule TestSubject do
  import BehaviorInjector

  define_hello()
end
```

The preceding code uses a macro defined in the `BehaviorInjector` module by simply invoking it in the `TestSubject` module. This alone is enough to define the `hello` function in `TestSubject`:

```
iex shell
iex> c "./behavior_injector.ex"
[BehaviorInjector]
iex> c "./test_subject.ex"
[TestSubject]
iex> TestSubject.hello()
Hello world!
:ok
```

We can see in the preceding code that using a macro makes injecting code a lot easier and cleaner. Another reason to use macros over `Code.eval_quoted/3` is that macros are more restrictive, and they don't allow you to pass environments and variables easily to manipulate the final results of a quoted literal evaluation.

The using macro

Elixir also has a special macro, `<module>.__using__/1`, which can be invoked by just calling the `use/2` keyword. This is a special macro that doesn't need us to explicitly call `require` with the module first before calling the macro.

For example, let's update `BehaviorInjector` to define a `__using__/1` macro:

behavior_injector.ex

```
defmodule BehaviorInjector do
  defmacro __using__(_options) do
    quote do
      def hello, do: IO.puts "Hello world!"
    end
  end
end
```

Now, in order to call that macro, all we need to do is use the `use/2` macro, as follows:

test_subject.ex

```
defmodule TestSubject do
  use BehaviorInjector
end
```

In the preceding code, we only passed one argument to the `use/2` macro because the second argument is optional. Now, we can see that the final result and the final behavior of `TestSubject` remained unchanged:

```
iex shell
iex> c "./behavior_injector.ex"
[BehaviorInjector]
iex> c "./test_subject.ex"
[TestSubject]
iex> TestSubject.hello()
Hello world!
:ok
```

Therefore, opt to define `__using__/1` macros to inject behavior into other modules, as they can be invoked by just calling `use/2`.

With this knowledge of injecting code using macros, let's take a look at one of the key ways to hook into a module's compilation and use these macros to change the behavior of a module, using compile-time callbacks.

Compile-time callbacks

Elixir also provides a mechanism to hook into the compilation of a module and perform tasks. These tasks can range from something as simple as logging to something as complex as injecting behavior.

These callbacks can be registered by using special reserved module attributes, but they are invoked depending on their type. There are three types of compile-time callbacks:

- `@before_compile`
- `@after_compile`
- `@on_definition`

Before compile

A callback registered using the `@before_compile` module attribute is registered as a before compile callback. As the name suggests, this callback is invoked right before the module is finished compiling.

Since this callback is invoked after all the functions in the module are defined, it has access to the module's functions and its attributes.

The callback can be either a function or a macro, but it needs to be defined in a different module outside of the current module, since it hasn't done compiling.

Here is an example of a before compile callback. We took the same code for the preceding `behavior_injector` and `test_subject` modules and used `@before_compile` to yield the same results:

behavior_injector.ex

```
defmodule BehaviorInjector do
  defmacro __before_compile__(_env) do
    quote do
      def hello, do: IO.puts "Hello world!"
    end
  end
end
```

In the preceding module, we replaced the __using__/1 macro with a __before_compile__/1 macro but kept the rest of the body identical. As you can see in the preceding code, __before_compile__/1 takes only one argument, and that is the environment in which the callback was registered:

test_subject.ex

```
defmodule TestSubject do
  @before_compile BehaviorInjector
end
```

We then updated TestSubject to add behavior, using the @before_compile callback instead of use/2. Since the name of the macro in BehaviorInjector was __before_compile__/1, all we needed to do was pass the module name to the @before_compile module attribute:

```
iex shell
iex> c "./behavior_injector.ex"
[BehaviorInjector]
iex> c "./test_subject.ex"
[TestSubject]
iex> TestSubject.hello()
Hello world!
:ok
```

Upon re-compiling the preceding files, we can see that the overall behavior of the TestSubject module remained the same.

After compile

A callback registered using the @after_compile module attribute is registered as an after compile callback. Just like its name, this callback is invoked right after the module is finished compiling.

Since the module is done compiling, this callback doesn't have access to unpersisted module attributes.

An after compile can either be a function or a macro, and it can also be defined in the same module itself.

Here is an example of an after compile callback:

test_after_compile.ex

```
defmodule TestAfterCompile do
  @after_compile __MODULE__

  def __after_compile__(_env, _bytecode) do
    IO.puts "Compiled #{__MODULE__}"
  end
end
```

In the preceding code, we registered an after compile callback to print a message when the module is finished compiling.

Therefore, upon compiling the preceding module in `iex`, we can see that the message is printed right after the compilation:

```
iex shell
iex> c "./test_after_compile.ex"
Compiled TestAfterCompile # prints compilation message
[TestAfterCompile]
```

The function or macro registered as an after compile callback takes the environment and the final bytecode of the compiled module.

On-definition

Callbacks registered using the `@on_definition` module attributes are on-definition callbacks. These are functions (not macros) that are invoked whenever a function or a macro is finished defining in the module.

Similar to `@before_compile`, this function needs to live in a different module, as the current module isn't done compiling.

In order to restrict what developers can do with this callback, Elixir only allows functions to be registered as on-definition callbacks.

Let's define a function that can be registered as an on-definition callback:

on_def.ex

```
defmodule OnDef do
  def __on_definition__(_, _, name, _, _, body) do
    IO.puts """
    Defining a function named #{name}
    with body:
    #{Macro.to_string(body)}
    """
  end
end
```

An on-definition callback function takes six arguments – the environment, the kind of function or macro (private or public), the name of the function or macro, a list of arguments as quoted literals, a list of guards as quoted literals, and a body of the function or macro as a quoted literal.

Here is an example of registering an on-definition callback:

test_on_def.ex

```
defmodule TestOnDef do
  @on_definition OnDef

  def hello, do: IO.puts "world"
end
```

In iex, we can see the message printed when compiling the test_on_def.ex file:

```
iex shell
iex> c "./on_def.ex"
[OnDef]
iex> c "./test_on_def.ex"
Defining a function named hello
with body:
[do: IO.puts("world")]
[TestOnDef]
```

The compile-time callbacks are invoked in the order in which they're registered. This means if a module has multiple @before_compile callbacks registered, it will invoke them in the order in which they were registered. This also means that future callbacks can override the functions/behaviors defined in the previous callbacks.

These three types of compile-time callbacks make Elixir a really powerful metaprogrammable language, with the ability to hook into a module's compilation and log/change its behavior. However, just like any metaprogramming construct, they should be used with a lot of thought and care.

Now that we understand some of the key constructs of metaprogramming in Elixir, let's use this to build a DSL in the next part.

Building a music DSL in Elixir

Now that we know the fundamentals of metaprogramming in Elixir, let's build a DSL. This DSL will be used to compose music using Elixir.

Here are the features of the DSL that we will be building:

- The DSL will provide a simple interface to compose music using Elixir
- It will utilize the ALSA aplay command to play a note
- The DSL should be able to define a sequence of notes so that the sequence can be reused throughout the track

- A note should have the following attributes:

  - A class: C, D, E, F, G, A, B, or rest

  - A modifier: sharp or base

  - A duration in seconds

  - A volume

  - An octet to determine the final frequency

- A sequence should be able to embed notes from another sequence

- The DSL should be easy to read, idiomatic, and deterministic

Now, let's come up with the shape of the DSL.

One of my favorite DSLs in Elixir is the Phoenix router. The interface is very simple and easy to learn, while still giving a lot of transparency to the developer using the DSL. So, taking inspiration from the Phoenix router, the following is the DSL I came up with for this project:

```
defmodule Music.DSLExample do
  use Music.DSL

  sequence :do_re_mi do
    note(:c, octet: 4, volume: 50, duration: 0.25)
    note(:d, modifier: :base, duration: 0.25)
    note(:e, modifier: :base, duration: 0.25)
  end

  sequence :fa_so_la_ti_do do
    note(:f, octet: 4, volume: 50, duration: 0.25)
    note(:g, modifier: :base, duration: 0.25)
    note(:a, duration: 0.25)
    note(:b, duration: 0.25)
    note(:c, octet: 5, duration: 0.25)
  end

  sequence :song do
    embed_notes(:do_re_mi)
    embed_notes(:fa_so_la_ti_do)
  end
end
```

The preceding code shows the DSL we will be building today. As mentioned previously, it's heavily inspired by the Phoenix router's pipeline and plug DSL.

Now, that we know what DSL we're building, let's work on the code for it. We can start by building the implementation (the logic behind the interface) first.

Writing the implementation

In this case, the implementation would be the modules used to define a note struct and play a note, using the `aplay` command.

> **Note for non-Linux readers**
>
> The `aplay` command only works for Linux machines with ALSA. If you're using a Mac or a Windows machine, you can try looking for alternatives to play a note as part of this project. Even if you cannot get a note to play on your machine, it would still be very useful to learn how to build this interface.

Since the goal of this project is to learn how to use metaprogramming, we will focus more of our attention on building the interface. Therefore, I will be glossing over some of the details of the implementation.

Let's start by creating a new mix project:

```
$ mix new mix_music
```

Now, inside our new mix project, let's define a `MixMusic.Note` module, which also defines a struct. This module will also be responsible for converting a note to a final frequency with a `MixMusic.Note.to_frequency/1` function:

ib/mix_music/note.ex

```
defmodule MixMusic.Note do
  @moduledoc """
  Struct representing a Music Note
  """

  defstruct class: nil, modifier: :base, octet: 4,
    duration: 0.5, volume: 50

  @a4_frequency 440
  @a4_octet 4
  @a4_index 9
  @frequency_constant 1.059463

  def to_frequency(%__MODULE__{class: :rest}), do: 0

  def to_frequency(%__MODULE__{octet: octet} = note) do
```

```
      semitones_from_a4 =
        12 * (octet - @a4_octet) - (@a4_index -
                                    semitone_index(note))
      relative_frequency = :math.pow(@frequency_constant,
                                     semitones_from_a4)
      round(@a4_frequency * relative_frequency)
    end

    defp semitone_index(%__MODULE__{class: class,
                                    modifier: modifier}) do
      Enum.find_index(
        semitones(),
        fn {cl, mo} -> class == cl && modifier == mo end
      )
    end

    defp semitones do
      [
        {:c, :base}, {:c, :sharp}, {:d, :base}, {:d, :sharp},
        {:e, :base},{:f, :base}, {:f, :sharp}, {:g, :base},
        {:g, :sharp}, {:a, :base},{:a, :sharp}, {:b, :base}
      ]
    end
  end
```

In the preceding module, we defined a `note` struct with fields for the class, modifier, octet, duration, and volume. We also defined the `to_frequency/1` function, which uses a mathematical equation to convert a note to a frequency using its semitones and octet.

Now, let's define a module to play a note that can also utilize the `to_frequency/1` function, defined in the preceding module:

lib/mix_music/note_player.ex

```
defmodule MixMusic.NotePlayer do
  @moduledoc """
  Plays a Note
  """
  alias MixMusic.Note
  require Logger

  def play(%Note{duration: duration, volume: volume} =
    note) do
    frequency = Note.to_frequency(note)
    Logger.info("Played Note: #{note.class}
```

```
                    #{note.modifier} #{note.octet}")

    cmd("""
    echo 'foo' |  awk '{ for (i = 0; i < #{duration};
       i+= 0.00003125) \
    printf("%08X\\n", #{volume}*sin(#{frequency}*3.14*
            exp((a[$1 % 8]/12)\
    *log(2))*i)) }' | xxd -r -p | aplay -c 2 -f S32_LE -r
    28000 """)
  end

  defp cmd(string) do
    case System.cmd("sh", ["-c", string]) do
      {_, 0} -> :ok
      _ -> :error
    end
  end
end
```

This module uses a terminal command to play a note, using its duration, volume, and frequency (calculated using the to_frequency/1 function). It relies on Linux's aplay command to play that note.

We have also added Logger.info/1 calls to help with testing. We will learn more about this later when we test this project.

Now that we have a way to define a note and play a note, we can define an interface to take a list of notes from the user and play them.

Building the interface layer

Let's define a new module, MixMusic.DSL, to define the DSL module. The first thing we need to do is convert a list of options to a Note struct. So, we will define a function to do that:

lib/mix_music/dsl.ex

```
defmodule MixMusic.DSL do
  @moduledoc """
  DSL to compose music in Elixir
  """

  def note_from_options(class, options) do
    params =
      options
      |> Keyword.put(:class, class)
```

```
      |> Enum.into(%{})

    struct!(MixMusic.Note, params)
  end
end
```

Next, let's start working on the interface part of this module by defining the __using__/1 macro. As part of the __using__/1 macro, we will do the following:

Import the MixMusic.DSL module because it will define macros and functions used in the DSL

Register a module attribute, sequences, to store a list of sequences while the module is still being defined

Define a function, play/1, which is responsible for playing a sequence of notes defined in the module:

lib/mix_music/dsl.ex

```
defmodule MixMusic.DSL do
  # ..
  defmacro __using__(_) do
    quote do
      import unquote(__MODULE__)

      Module.register_attribute(__MODULE__, :sequences,
                                accumulate: true)

      def play(sequence_name) do
        # TODO
      end
    end
  end
end
```

Now, let's define the sequence/2 macro, which is used to define and collect a list of notes. This macro will be responsible for the following:

- Registering a module attribute, current_sequence, indicating that a sequence is being defined.
- Unquoting the given block, which would update the current_sequence module attribute.

Adding the notes present in the current_sequence attribute to the sequences attribute registered in the __using__/1 macro. The list of notes in current_sequence needs to be reversed because the module attribute accumulates by adding new elements to the beginning of the list.

Deleting the `current_sequence` module attribute, indicating that we've finished defining the current sequence:

lib/mix_music/dsl.ex

```
defmodule MixMusic.DSL do
  # ..
  defmacro sequence(sequence_name, do: block) do
    quote do
      Module.register_attribute(__MODULE__,
        :current_sequence, accumulate: true)
      unquote(block)
      @sequences {unquote(sequence_name),
                  @current_sequence |> Enum.reverse()}
      Module.delete_attribute(__MODULE__,
                              :current_sequence)
    end
  end
end
```

Now that we have a way to define a sequence, let's define the `note/2` macro, which will be responsible for updating the current sequence. The `note/2` macro will be responsible for the following:

- Checking whether a sequence is currently being defined by checking the `@current_sequence` module attribute. This is important because we want to restrict calling the `note/2` macro to inside the `sequence/2` macro. This is consistent with what we learned earlier in the chapter about making a DSL as restrictive as possible, in order to make its behavior more deterministic.

Using the `note_from_options/2` function defined previously to add the new note to the `@ current_sequence` attribute:

lib/mix_music/dsl.ex

```
defmodule MixMusic.DSL do
  # ...
  defmacro note(class, options) do
    quote do
      if @current_sequence do
        @current_sequence note_from_options(unquote(class),
          unquote(options))
      else
        raise "note can only be called inside a sequence"
      end
    end
```

```
      end
   end
```

Similar to the note/2 macro, we also need the embed_notes/1 macro, which will be used to embed notes from one sequence to another. The embed_notes/1 macro will be responsible for the following:

Checking whether @current_sequence is defined (similar to note/2)

Fetching a list of notes from an already defined sequence and adding each of them to the @current_sequence attribute:

lib/mix_music/dsl.ex

```elixir
defmodule MixMusic.DSL do
  # ..
  defmacro embed_notes(sequence_name) do
    quote do
      if @current_sequence do
        notes = notes_from_sequence(__MODULE__,
                                    unquote(sequence_name))

        for note <- notes do
          @current_sequence note
        end
      else
        raise "embed_notes can only be called inside a
               sequence"
      end
    end
  end

  def notes_from_sequence(mod, sequence_name) do
    mod
    |> Module.get_attribute(:sequences)
    |> Keyword.get(sequence_name)
  end
end
```

We also defined a helper function, notes_from_sequence/2, to get notes from an already defined sequence for a module while it's still being compiled.

Now, we have a list of sequences defined for any module in the `@sequences` module attribute. We can use a `@before_compile` hook to utilize that module attribute, to define a function at the end of the module's compilation:

lib/mix_music/dsl.ex

```
defmodule MixMusic.DSL do
  # ..
  defmacro __using__(_) do
    quote do
      @before_compile unquote(__MODULE__)
      import unquote(__MODULE__)

      Module.register_attribute(__MODULE__, :sequences,
                                accumulate: true)

      def play(sequence_name) do
        # TODO
      end
    end
  end
  # ..
  defmacro __before_compile__(_) do
    quote do
      def __sequences__, do: @sequences
    end
  end
end
```

Now, at the end of the compilation of a module using `MixMusic.DSL`, there will be a `__sequences__/0` function defined, which returns a `Keyword` list of sequence names that has a list of notes, for all the sequences in the module. Defining this function also increases the introspective features of this DSL.

Now that we have a way to get a list of notes for a sequence, we can define a way to play a sequence for a module. This function will use the `MixMusic.NotePlayer.play/1` function defined previously to do that. We will also update the `play/1` function defined inside the `__using__/1` macro:

lib/mix_music/dsl.ex

```
defmodule MixMusic.DSL do
  # ..
  defmacro __using__(_) do
    quote do
```

```elixir
      @before_compile unquote(__MODULE__)
      import unquote(__MODULE__)

      Module.register_attribute(__MODULE__, :sequences,
                                 accumulate: true)

      def play(sequence_name) do
        play_sequence(__MODULE__, sequence_name)
      end
    end
  end
  # ..
  def play_sequence(mod, sequence_name) do
    case Keyword.get(mod.__sequences__, sequence_name) do
      nil -> raise "sequence #{sequence_name} not defined"
      notes -> Enum.each(notes,
                         &MixMusic.NotePlayer.play/1)
    end
  end
end
```

With the preceding code, we can now play any sequence defined in a module. Now, our DSL can be used as follows:

lib/mix_music/dsl_example.ex

```elixir
defmodule MixMusic.DSLExample do
  use MixMusic.DSL

  sequence :do_re_mi do
    note(:c, octet: 4, volume: 50, duration: 0.25)
    note(:d, modifier: :base, duration: 0.25)
    note(:e, modifier: :base, duration: 0.25)
  end

  sequence :fa_so_la_ti_do do
    note(:f, octet: 4, volume: 50, duration: 0.25)
    note(:g, modifier: :base, duration: 0.25)
    note(:a, duration: 0.25)
    note(:b, duration: 0.25)
    note(:c, octet: 5, duration: 0.25)
  end

  sequence :song do
    embed_notes(:do_re_mi)
```

```
        embed_notes(:fa_so_la_ti_do)
    end
end
```

We can now start an `iex` console and play the preceding module's `song` sequence:

```
$ iex -S mix
iex> MixMusic.DSLExample.play(:song)
...
... aplay output
:ok
```

Now that we have the DSL ready, it's time to test it.

Testing the interface

We have a few things to test in this DSL:

- **Integration**: Testing whether the DSL plays well with the `NotePlayer` module to ensure the whole project is functional

- **Sequences**: Testing whether sequences are defined in the correct order and the defaults for `note` are working

- **Restrictions**: Testing whether the DSL works as expected and `note` and `embed_notes` can only be called inside `sequence`

Writing integration tests

Let's start by adding the integration test:

---

test/mix_music/integration_test.exs

```
defmodule MixMusic.DoReMi do
  use MixMusic.DSL

  sequence :do_re_mi do
    note(:c, octet: 4, volume: 50, duration: 0.25)
    note(:d, modifier: :base, duration: 0.25)
    note(:e, modifier: :base, duration: 0.25)
  end

  sequence :fa_so_la_ti_do do
    note(:f, octet: 4, volume: 50, duration: 0.25)
    note(:g, modifier: :base, duration: 0.25)
```

```
      note(:a, duration: 0.25)
      note(:b, duration: 0.25)
      note(:c, octet: 5, duration: 0.25)
    end

    sequence :song do
      embed_notes(:do_re_mi)
      embed_notes(:fa_so_la_ti_do)
    end
  end

  defmodule MixMusic.IntegrationTest do
    use ExUnit.Case
    import ExUnit.CaptureLog

    describe "DoReMi.play/1" do
      test "plays the song (not testing alsa)" do
        output = capture_log(fn ->
                             MixMusic.DoReMi.play(:song) end)

        assert output =~ "Played Note: c base 4"
        assert output =~ "Played Note: d base 4"
        assert output =~ "Played Note: e base 4"
        assert output =~ "Played Note: f base 4"
        assert output =~ "Played Note: g base 4"
        assert output =~ "Played Note: a base 4"
        assert output =~ "Played Note: b base 4"
        assert output =~ "Played Note: c base 5"
      end
    end
  end
```

Since there is no easy way of testing the `aplay` commands, we use the `Logger` calls inside the `NotePlayer.play/1` function to ensure proper notes are being played. Another way of testing this is to use `Mox`, which would serve a very similar purpose.

Testing sequence

Now, it's time to test the `sequence/2` macro and whether it updates the `@sequences` attribute:

test/mix_music/dsl/sequence_test.exs

```
defmodule QuotedModule.Sequence do
  Module.register_attribute(__MODULE__, :sequences,
                           accumulate: true)
```

```
  def sequences_before, do: @sequences

  require MixMusic.DSL

  MixMusic.DSL.sequence :test do
    @current_sequence 5
    @current_sequence 10
  end

  def sequences_after, do: @sequences
end

defmodule MixMusic.DSL.SequenceTest do
  use ExUnit.Case

  describe "sequence/2" do
    test "updates `@sequences`" do
      assert QuotedModule.Sequence.sequences_before() == []
      assert QuotedModule.Sequence.sequences_after() ==
        [test: [5, 10]]
    end
  end
end
```

We can test the sequence/2 macro in isolation by testing whether calling it updates the @sequences module attribute, as expected. We can define a sequences_before/0 function, which stores the state of the module attribute before calling sequence/2. Similarly, we can define a sequences_after/0 function, which stores the state of the @sequences attribute after the sequence/2 call. Then, we can test to ensure that sequences_before/0 is [] and sequences_after/0 is the updated state of @sequences, as expected. This way, we can ensure that the sequence/2 macro works as expected in isolation with note/2 and the embed_notes/1 macro.

Testing note and embed_notes

In order to test whether note and embed_notes work well with sequence/2, it's best to define a module that uses sequence as expected.

Let's first define a support module to test the DSL:

test/support/mix_music/dsl_tester.ex

```
defmodule MixMusic.DSLTester do
  use MixMusic.DSL
```

```
  sequence :do_re_mi do
    note(:c, octet: 4, volume: 50, duration: 0.25)
    note(:d, modifier: :base, duration: 0.25)
    note(:e, modifier: :base, duration: 0.25)
  end

  sequence :fa_so_la_ti_do do
    note(:f, octet: 4, volume: 50, duration: 0.25)
    note(:g, modifier: :base, duration: 0.25)
    note(:a, duration: 0.25)
    note(:b, duration: 0.25)
    note(:c, octet: 5, duration: 0.25)
  end

  sequence :song do
    embed_notes(:do_re_mi)
    embed_notes(:fa_so_la_ti_do)
  end
end
```

We can now write a test file that utilizes the preceding `DSLTester` module. We will write a test to ensure the following:

- `play_sequence/2` works as expected for a valid sequence

- `play_sequence/2` raises an error for an invalid sequence

- `__before_compile__/1` defines a `__sequences__/0` function that returns the state of the `@sequences` module attribute:

test/mix_music/dsl_test.exs

```
Code.require_file("test/support/mix_music/dsl/tester.ex")

defmodule MixMusic.DSLTest do
  use ExUnit.Case
  import ExUnit.CaptureLog

  alias MixMusic.{DSL, DSLTester}

  describe "play_sequence/2" do
    test "raises when calling a sequence that's not
          defined" do
      assert_raise(RuntimeError,
        ~r/sequence bad_sequence not defined/, fn ->
        DSLTester.play(:bad_sequence)
```

```
        end)
    end

    test "plays a sequence when it's defined" do
      output = capture_log(fn -> DSLTester.play(:do_re_mi)
                           end)

        assert output =~ "Played Note: c base 4"
        assert output =~ "Played Note: d base 4"
        assert output =~ "Played Note: e base 4"
    end
  end

  describe "__before_compile__/1" do
    require DSL

    test "defines a function `__sequences__` which
          delegates to `@sequences`" do
      expected =
        quote do
          def __sequences__, do: @sequences
        end

      expr = quote do: DSL.__before_compile__(nil)
      expanded = Macro.expand_once(expr, __ENV__)

      assert Macro.to_string(expected) ==
        Macro.to_string(expanded)
    end

    test "defines `DSLTester.__sequences__/0`" do
      sequences = Keyword.keys(DSLTester.__sequences__())

      assert :song in sequences
      assert :do_re_mi in sequences
      assert :fa_so_la_ti_do in sequences
    end
  end
end
```

The preceding test utilizes tools like `Macro.to_string/1` to compare the quoted literal defined in a macro with the expected quoted literal, which defines a `__sequences__/1` function. I have found this technique extremely helpful to test macros in smaller chunks without having to test the entire DSL.

We also used the `Logger` calls again in the `NotePlayer` module to test the `play_sequence/1` function.

Testing restrictions

In order to ensure that our DSL is designed well, we should make sure that it restricts developers from misusing parts of the DSL. In our case, one example is not allowing the usage of `note` and `embed_notes` outside of `sequence`.

So, let's write a test for that:

test/mix_music/dsl/restrictions_test.exs

```elixir
defmodule MixMusic.DSL.RestrictionsTest do
  use ExUnit.Case

  describe "note/2" do
    test "can only be called inside a sequence" do
      quoted_module =
        quote location: :keep do
          defmodule QuotedModule.Note do
            import MixMusic.DSL

            note(:c, volume: 10)
          end
        end

      assert_raise(RuntimeError, "note can only be called
        inside a sequence", fn ->
        Code.eval_quoted(quoted_module)
      end)
    end
  end

  describe "embed_notes/2" do
    test "can only be called inside a sequence" do
      quoted_module =
        quote location: :keep do
          defmodule QuotedModule.Note do
            import MixMusic.DSL

            embed_notes(:sequence)
          end
        end
```

```
        assert_raise(RuntimeError, "embed_notes can only be
          called inside a sequence", fn ->
          Code.eval_quoted(quoted_module)
        end)
      end
    end
  end
```

In the preceding test file, we simply defined quoted literals where modules called the note/2 and embed_notes/1 macros outside of a sequence/2 macro, and expected RuntimeError to be raised when evaluating those quoted literals.

This is another use case for quoted literals, where you can separate defining the code of a module from actually defining the module by evaluating the quoted literals.

There are still many tests that we could add to increase the determinism of this DSL, but in order to save us some time, we will not be exploring them.

We now know how to test quoted literals generated by macros, but it was a lot of work to test the generated quoted literals using Code.eval_quoted/2. Therefore, one of the key features to add to a DSL is helper functions that make testing the DSL easier for developers using it. Let's add test helpers to our DSL in the next module.

Adding DSL test helpers

One key element to making your DSL easy to use is to add easier ways to test the DSL. Oftentimes, this means adding ways to test whether the DSL is being used correctly and will yield the expected output. Most of this can be done by adding some introspective features, such as the __sequences__/0 function.

In order to make our DSL more testable, we can define a cleaner way to test whether a sequence was defined with expected notes. Let's define the MixMusic.DSLTestHelper module:

lib/mix_music/dsl/test_helpers.ex

```
defmodule MixMusic.DSL.TestHelper do
  @moduledoc """
  Test helpers for MixMusic.DSL
  """
  def defines_sequence?(module, sequence_name,
                        with_notes: notes) do
    actual_notes =
      module.__sequences__()
      |> Keyword.get(sequence_name)

    notes == actual_notes
```

```
    end

    def sequence_note(class, options) do
      MixMusic.DSL.note_from_options(class, options)
    end
  end
```

In the preceding module, we can use the `defines_sequence?` function to test whether a module defines a sequence with a list of notes. In order to keep the DSL simple, I have also defined a `sequence_note/2` function, which can be used to test whether a list of notes is in sequence.

Here is an example of using the `DSL.TestHelper` module to test the `MixMusic.DSLTester` module defined earlier:

test/mix_music/dsl/test_helpers_test.exs

```
Code.require_file("test/support/mix_music/dsl/tester.ex")

defmodule MixMusic.DSL.TestHelperTest do
  use ExUnit.Case

  import MixMusic.DSL.TestHelper

  describe "test DSL" do
    test "defines do_re_mi with correct notes" do
      assert defines_sequence?(
        MixMusic.DSLTester,
        :do_re_mi,
        with_notes: [
          sequence_note(:c, octet: 4, volume: 50,
                        duration: 0.25),
          sequence_note(:d, modifier: :base,
                        duration: 0.25),
          sequence_note(:e, modifier: :base,
                        duration: 0.25)
        ]
      )
    end
  end
end
```

This way, we can more easily test a module using `MixMusic.DSL`, without having to write an integration test or use `Mox`.

Summary

In this chapter, we started by learning what metaprogramming is and its pros and cons. We also learned that metaprogramming should be used carefully, as it makes code more complex, and outlined cases when it is okay to leverage metaprogramming.

We learned that in Elixir, metaprogramming revolves around three constructs – quoted literals, macros, and compile-time callbacks. We saw how quoted literals are Elixir's representation of an AST, which facilitates most metaprogramming. We then learned how to use macros to inject behavior, using quoted literals in a module at compile time. We also learned that quoted literals and macros are evaluated hygienically, but `var!/2` can be used to bypass the hygiene and define variables outside of the scope of a quote. We then took a look at compile-time callbacks, which are used to run tasks (or add behavior) by hooking into the compilation of a module.

We then proceeded to see how a DSL should be designed and that it should be idiomatic, introspective, testable, and clearly separated from the implementation. Finally, we built a functional and fully tested DSL to build music using Elixir.

In the next chapter, we will use the concepts of metaprogramming covered in this chapter to build a DSL for our controller's interface and the view interface from scratch.

Exercises

One of the ways we can improve the DSL is to add better integration tests. I previously mentioned using Mox, but can you think of any other ways to test the DSL using some of the metaprogramming concepts that we have learned in this chapter?

9
Controller and View DSL

In this chapter, we will use the metaprogramming skills learned in the previous chapter to make our controller and view modules easier to read and extend. We will first define the requirements for our interface, understand why we need metaprogramming, and build an idiomatic interface to properly define future controllers and views.

We will cover the following topics in this chapter:

- Understanding why it makes sense to use metaprogramming here
- Designing the DSL for controllers based on requirements
- Using concepts covered in the previous chapter to build a DSL for controllers
- Adding reflections to make it easier to debug and test the controller DSL
- Making `plug` calls more introspective and reflective
- Designing and building the DSL for views
- Adding reflections to make it easier to debug and test the view DSL
- Building an app to test out the new DSLs
- Creating test helper modules to make it easier to test the DSLs

At the end of this chapter, we will have a functioning **domain-specific language** (**DSL**), which will simplify defining controllers and views when used with the Goldcrest web framework.

Technical requirements

In this chapter, we will return to the example web application we built in *Part 2* of this book. We will look at the controllers and views in those applications and use those to define the requirements of our DSLs.

Similar to previous chapters, Elixir `1.12.x` and Erlang `23.2.x` are required to properly run the code snippets in this chapter. I'd also recommend coding along and understanding why we decided to use metaprogramming for this use case.

The code examples for this chapter can be found at `https://github.com/PacktPublishing/Build-Your-Own-Web-Framework-in-Elixir/tree/main/chapter_09`

Why use metaprogramming here?

Based on what we learned in *Chapter 8*, we should use metaprogramming only after trying to implement our solution without using any metaprogramming. We spent *Chapter 1* to *Chapter 7* learning how to build a web framework without metaprogramming, so now it's a good time to consider whether or not we should use it.

Here are the reasons why we should use metaprogramming for `Goldcrest`:

- We will be building DSLs that will be used in multiple places, by multiple controllers that use `Goldcrest`. This will allow us to standardize how controllers and views behave in `Goldcrest`.

- We have a clear set of requirements that we don't anticipate changing very often. This allows us to build a stable DSL for the `Goldcrest` controllers and views.

- Our implementation without a DSL is too complex to read when developers just need to make simple changes to define a controller. That's a clear indication that this is a great candidate to build a DSL to wrap around the complexities of the implementation.

Now that we have understood why a DSL in conjunction with metaprogramming will be a good fit here, let's take a look at the final design of the DSL we want to build.

DSL design

For our DSL to be readable and easier to digest, it must show only the information that's needed in our controller. Let's take the following controller as an example. This is how we would write a controller today for it to work well with `Goldcrest`:

```
defmodule TasksWeb.TaskController do
  use Plug.Builder

  import Goldcrest.Controller
  import Plug.Conn

  alias TasksWeb.Tasks

  plug :ensure_authorized!

  def call(conn, action: action) do
```

```
    conn = super(conn, [])

    apply(__MODULE__, action, [conn, conn.params])
  end

  def index(conn, _params) do
    tasks = Tasks.list()

    conn
    |> put_status(200)
    |> render("tasks.html.eex", tasks: tasks)
  end

  def create(conn, %{"name" => name, "description" =>
    description}) do
    Tasks.add(name, description)

    conn
    |> redirect(to: "/tasks")
  end

  def delete(conn, %{"id" => id}) do
    id
    |> String.to_integer()
    |> Tasks.delete()

    conn
    |> redirect(to: "/tasks")
  end

  # ..
  defp render(conn, file, assigns) do
    TasksWeb.TaskView.render(conn, file, assigns)
  end
end
```

In the preceding controller, we need the use `Plug.Builder` line so we can invoke the `plug/2` macro, which provides a simple way to call function plugs before every request. We also need `import Plug.Conn`, as most of the time, our controllers will be manipulating a `Plug.Conn` struct and the `Plug.Conn` module provides several functions that can be used to accomplish that. Finally, we also need the `call/2` function because that's what gets called from our router as part of the match and dispatch process. Therefore, a good final interface for our controller would be removing those items and substituting them with a `use` statement. We can also remove the `render/3` function at the bottom and define that within the `use` statement to increase readability:

```
defmodule TasksWeb.TaskController do
  use Goldcrest.Controller
  alias TasksWeb.Tasks

  plug :ensure_authorized!

  def index(conn, _params) do
    tasks = Tasks.list()

    conn
    |> put_status(200)
    |> render("tasks.html.eex", tasks: tasks)
  end

  def create(conn, %{"name" => name, "description" => description}) do
    Tasks.add(name, description)

    conn
    |> redirect(to: "/tasks")
  end

  def delete(conn, %{"id" => id}) do
    id
    |> String.to_integer()
    |> Tasks.delete()

    conn
    |> redirect(to: "/tasks")
  end
end
```

In the preceding code, we have restricted the code that adds the controller behavior to our module to a single line, use `Goldcrest.Controller`. This makes the `TaskController` module easier to read and maintain while enforcing a standard pattern of creating controller modules.

Building the controller DSL

To successfully call the use Goldcrest.Controller statement, we will need to define a __using__/1 macro in Goldcrest.Controller, which is responsible for injecting a controller-like behavior into a module.

Let's start by simply moving the Plug.Builder call, the Plug.Conn call, and the call/2 function to the Goldcrest.Controller.__using__/1 macro:

```
defmodule Goldcrest.Controller do
  import Plug.Conn

  defmacro __using__(_) do
    quote do
      use Plug.Builder
      import Plug.Conn

      def call(conn, action: action) do
        conn = super(conn, [])

        apply(__MODULE__, action, [conn, conn.params])
      end
    end
  end

  # ..
end
```

In the previous code snippet, we've added a new __using__/1 macro, which defines how a controller should look in Goldcrest.

There's one more piece of controller code that we need to extract, the render/3 function. This one's a little tricky because the render/3 function calls the render/3 function defined in the view module, which is similarly named to the controller module. Therefore, we will need to do some string jujitsu to get the view name. We'll keep it simple for now and add the following code to the controller:

```
defmodule Goldcrest.Controller do
  defmacro __using__(opts) do
    quote do
      # ..
      @opts unquote(opts)

      @default_view_module __MODULE__
                           |> Module.split()
                           |> List.update_at(-1, fn str ->
                             String.replace(str,
```

```
                                "Controller", "View")
                    end)
                    |> Module.concat()

      @view_module @opts[:view_module] ||
        @default_view_module

      def render(conn, file, assigns) do
        @view_module.render(conn, file, assigns)
      end

      def call(conn, action: action) do
        conn = super(conn, [])

        apply(__MODULE__, action, [conn, conn.params])
      end
    end
  end

  # ..
end
```

We used the current module name in the preceding code snippet to get the corresponding `View` module name. We first use `Module.split/1` to split the controller module's name by namespaces. We then replace the `Controller` substring in the last element of the split list with `View` to get the default view name. Since it's possible that the view name doesn't follow the same naming conventions as the controller modules, we also allow it to be overridden by using the optional argument with the `use` call. This allows us to have a reasonable default for the view module name while still giving developers the ability to override it. We then define the `render/3` function, which further delegates to the view module.

Now that we have added a base structure for defining a controller, let's add some reflections to help us better test and inspect our code.

Adding reflections

In this section, we will add reflections to `Goldcrest.Controller`. These functions are similar to the introspective `__sequences__/0` function we added in *Chapter 8*, where they will allow us to test, debug, and inspect our DSL better.

Let's start by adding functions to return the options that use `Goldcrest.Controller` was called with:

```
defmodule Goldcrest.Controller do
  defmacro __using__(opts) do
    quote do
      # ..
      @opts unquote(opts)

      @default_view_module __MODULE__
                           |> Module.split()
                           |> List.update_at(-1, fn str ->
                             String.replace(str,
                               "Controller", "View")
                           end)
                           |> Module.concat()

      @view_module @opts[:view_module] ||
        @default_view_module

      def render(conn, file, assigns) do
        @view_module.render(conn, file, assigns)
      end

      def __goldcrest_controller_using_options__, do: @opts

      def call(conn, action: action) do
        conn = super(conn, [])

        apply(__MODULE__, action, [conn, conn.params])
      end
    end
  end

  # ..
end
```

In the preceding code, we have defined a `__goldcrest_controller_using_options__/0` function that returns the compile-time value of the @opts module attribute. As we can see, @opts gets assigned using the unquote/1 macro to the opts value passed to the `__using__`/1 macro. In this way, even after compilation, we can easily tell what options were passed to the use Goldcrest.Controller call, by simply calling the `__goldcrest_controller_using_options__`/0 function.

> **Naming Reflection functions**
>
> Traditionally, reflection functions are named with surrounding __ (two underscores). This makes it such that the iex completion helpers ignore these functions, making it very clear that their intended use is within the library that's defining them. I like to think of the two underscores as a reflective surface of sorts, so I think it's a pretty good convention for reflective functions. I'm not sure whether that was intended by the Elixir core team.
>
> I also like to be very explicit while naming my reflections. Instead of just naming them __ options__, I prefer __controller_using_options__. This significantly decreases the likelihood of collisions and makes it clear what the reflection is about and which library it is being used for.

Similarly, we will define a function to return the `view_module` that represents the module of the controller. This can simply be the `view_module` passed along in the `opts` or `default_view_module` based on the controller module name:

```
defmodule Goldcrest.Controller do
  defmacro __using__(opts) do
    quote do
      # ..
      @opts unquote(opts)

      @default_view_module __MODULE__
                           |> Module.split()
                           |> List.update_at(-1, fn str ->
                             String.replace(str,
                               "Controller", "View")
                           end)
                           |> Module.concat()

      @view_module @opts[:view_module] ||
        @default_view_module

      def render(conn, file, assigns) do
        @view_module.render(conn, file, assigns)
      end

      def __goldcrest_controller_using_options__, do: @opts

      def __goldcrest_controller_view_module__, do:
        @view_module

      def call(conn, action: action) do
        conn = super(conn, [])

        apply(__MODULE__, action, [conn, conn.params])
```

```
        end
      end
    end

    # ..
end
```

We can now tell at runtime which module or view the render function is being delegated to. Now that we have added some of the simpler reflections, let's take a look at one of the more complex ones that allow us to inspect all the plug/2 calls.

Reflective plugs

One of the tricky things to test in a controller is all the combinations of plugs registered using the plug/2 call. They might interfere with each other, and since testing them is hard, developers often prefer not to test them properly. On top of that, it's hard to know what plugs will be called and in what order they will be called at runtime. One way of fixing both of these problems is by adding reflective functions.

In this section, we will learn how to use metaprogramming tools to add a reflective function to fetch all the plugs registered using the plug/2 call in a controller's compile-time context.

Let's start by returning to *Chapter 5*, where we learned the Plug.Builder module stores the list of all the registered plugs that were registered using the plug/2 macro in a @plugs collectible module attribute. This means that when the final call/2 function is invoked, it will have access to all the plugs using the @plugs module attribute. If we could access the attribute at runtime, that would solve our problem, but the way Plug.Builder defines the attribute, it's not accessible after compilation.

If we were to add a __goldcrest_controller_plugs__/0 function at the very end of the controller module, it would work:

```
defmodule TasksWeb.TaskController do
  # ..
  def __goldcrest_controller_plugs__, do: @plugs
end
```

The preceding code snippet allows us to register all the plugs using the plug/2 call by simply calling __goldcrest_controller_plugs__/0. This works because the function is defined at compile time when the attribute is accessible and because it is added to the very end of the module. Adding this function to the end of the controller module allows it access to all the plug/2 calls made before it.

Keeping this in mind, let's now try adding this to the Goldcrest.Controller.__using__/1 macro:

```
defmodule Goldcrest.Controller do
  defmacro __using__(opts) do
    quote do
```

```
        @opts unquote(opts)

        @default_view_module __MODULE__
                        |> Module.split()
                        |> List.update_at(-1, fn str ->
                          String.replace(str,
                            "Controller", "View")
                        end)
                        |> Module.concat()

        @view_module @opts[:view_module] ||
          @default_view_module

        def render(conn, file, assigns) do
          @view_module.render(conn, file, assigns)
        end

        def __goldcrest_controller_using_options__, do: @opts

        def __goldcrest_controller_view_module__, do:
          @view_module

        def __goldcrest_controller_plugs__, do: @plugs

        def call(conn, action: action) do
          conn = super(conn, [])

          apply(__MODULE__, action, [conn, conn.params])
        end
      end
    end

  # ..
end
```

In the previous code snippet, we defined the `__goldcrest_controller_plugs__`/0 function at the end of the `__using__`/1 macro, not at the end of the compilation of the module. Now, let's take the following controller module as an example:

```
defmodule ExampleController do
  use Goldcrest.Controller

  plug :test_plug
  plug :test_plug2
```

```
    #  ..
  end
```

In the preceding code, since `__goldcrest_controller_plugs__`/0 gets defined on line 2 as part of the `use Goldcrest.Controller` call, it never captures the updated @plugs attribute after line 4 and line 5, where two new plugs are added to the pipeline. Therefore, we need to ensure that the function is not just defined at the end of the `__using__`/1 macro, but also at the end of the module right before the module gets compiled. This is the perfect candidate for a `@before_compile` hook, as we learned in *Chapter 8*.

Let's start by defining a helper module that houses the `before_compile` macro. Remember, the `before_compile` macro cannot be defined in the same module in which it's registered:

```
defmodule Goldcrest.Controller.BeforeCompileHelpers do
  defmacro __before_compile__(_macro_env) do
    quote do
      def __goldcrest_controller_plugs__, do: @plugs
    end
  end
end
```

The preceding module defines a macro, which on invoking, defines the `__goldcrest_controller_plugs__`/0 function. Now, all that's left is registering this macro as a `@before_compile` callback in `Goldcrest.Controller`:

```
defmodule Goldcrest.Controller do
  defmacro __using__(opts) do
    quote do
      @opts unquote(opts)

      @default_view_module __MODULE__
                           |> Module.split()
                           |> List.update_at(-1, fn str ->
                             String.replace(str,
                               "Controller", "View")
                           end)
                           |> Module.concat()

      @view_module @opts[:view_module] ||
        @default_view_module

      def render(conn, file, assigns) do
        @view_module.render(conn, file, assigns)
      end

      def __goldcrest_controller_using_options__, do: @opts
```

```
        def __goldcrest_controller_view_module__, do:
          @view_module

        def call(conn, action: action) do
          conn = super(conn, [])

          apply(__MODULE__, action, [conn, conn.params])
        end

        @before_compile
          Goldcrest.Controller.BeforeCompileHelpers
      end
    end

    # ..
  end
```

In the preceding code, we added a @before_compile callback, which gets called at the very end of the controller's compilation. This allows it to register all the plug/2 calls and capture the final state of the @plugs module attribute.

Now that we have the controller DSL done, it's time to work on building a simpler interface for views.

Extracting response helpers

First, let's move the render functions in Goldcrest.View to a helper module, which can also be imported into the Goldcrest.Controller.__using__/1 macro, allowing the controller to use functions such as redirect and render:

```
defmodule Goldcrest.ResponseHelpers do
  @moduledoc false

  import Plug.Conn
  require EEx

  def render(conn, :json, data) when is_map(data) do
    conn
    |> content_type("application/json")
    |> send_resp(200, Jason.encode!(data))
  end

  def render(conn, :json, data) when is_binary(data) do
    data = ensure_json!(data)
```

```elixir
  conn
  |> content_type("application/json")
  |> send_resp(conn.status || 200, data)
end

def render(conn, :html, data) when is_binary(data) do
  conn
  |> content_type("text/html")
  |> send_resp(200, data)
end

def render(conn, file, opts) do
  contents =
    file
    |> html_file_path()
    |> EEx.eval_file(opts)

  render(conn, :html, contents)
end

defp html_file_path(file) do
  templates_path()
  |> Path.join(file)
end

defp templates_path do
  Application.fetch_env!(:goldcrest, :templates_path)
end

def content_type(conn, content_type) do
  put_resp_content_type(conn, content_type)
end

def redirect(conn, to: url) do
  body = redirection_body(url)

  conn
  |> put_resp_header("location", url)
  |> content_type("text/html")
  |> send_resp(conn.status || 302, body)
end

defp redirection_body(url) do
  html = Plug.HTML.html_escape(url)
```

```
    "<html><body>You are being <a href=\"#{html}\">
      redirected</a>" <>
      ".</body></html>"
  end

  defp ensure_json!(data) do
    data
    |> Jason.decode!()
    |> Jason.encode!()
  end
end
```

In the preceding code snippet, we moved the render/3 function, the redirect/2 function, and the content_type/2 function to a new Goldcrest.ResponseHelpers helper module. We also used the Jason.encode!/1 and Jason.decode!/2 functions to define a clause for the render/3 function, which handles the application/json type request.

We also handle clauses where we can send either a map or a string to a json rendering clause, and Goldcrest.ResponseHelpers handles both cases. For a map, Goldcrest.ResponseHelpers uses Jason.encode!/1 to send a JSON response. In this way, anything inside a map with the Jason.Encoder protocol implemented can be sent to Goldcrest.ResponseHelpers as part of the response body. When a string is sent to Goldcrest.ResponseHelpers, it first ensures that the string is a valid json response by piping the raw string to Jason.decode!/1 and then Jason.encode!/1. If this function doesn't raise, it simply sends the raw string as the body of the response.

Now that we have a central place to define all response-related functions, we can update Goldcrest.Controller to import it. Doing this really simplifies the Goldcrest.Controller module:

```
defmodule Goldcrest.Controller do
  import Plug.Conn

  defmacro __using__(opts) do
    quote do
      use Plug.Builder
      import Plug.Conn

      import Goldcrest.ResponseHelpers, except: [render: 3]

      @opts unquote(opts)

      @default_view_module __MODULE__
                           |> Module.split()
                           |> List.update_at(-1, fn str ->
                             String.replace(str,
```

```
                            "Controller", "View")
                    end)
                    |> Module.concat()

    @view_module @opts[:view_module] ||
      @default_view_module

    def render(conn, file, assigns) do
      @view_module.render(conn, file, assigns)
    end

    def __goldcrest_controller_using_options__, do: @opts

    def __goldcrest_controller_view_module__, do:
      @view_module

    @before_compile
      Goldcrest.Controller.BeforeCompileHelpers

    def call(conn, action: action) do
      conn = super(conn, [])

      apply(__MODULE__, action, [conn, conn.params])
    end
  end
 end
end
```

In the previous code snippet, we added the import Goldcrest.ResponseHelpers call, but we're not importing render/3 from Goldcrest.ResponseHelpers. This is because we want to go through the View layer for rendering. This will allow us to define rendering-related helper functions in the View module, keeping the View module completely responsible for any render/3 calls. Now that we have extracted response helper functions and simplified the Goldcrest.Controller module, it's time to build our View DSL.

Building the View DSL

Let's first start by understanding what our view modules should look like. Let's take a look at the following view module:

```
defmodule TasksWeb.TaskView do
  use Goldcrest.View

  def stringify_task(_task = {name, description}) do
```

```
      "#{name} - #{description}"
    end
end
```

In the preceding module, we have defined a helper function that converts a task to a string, which can be further used in a template as shown in the following code snippet:

```
<h1>Listing Tasks</h1>
<table>
  <thead>
    <tr>
      <th>Task</th>
      <th></th>
    </tr>
  </thead>
  <tbody>
  <%= for {task, index} <- Enum.with_index(@tasks) do %>
    <tr>
      <td><%= stringify_task(task) %></td>
      <td>
        <a href="/tasks/<%= index %>/delete">Delete</a>
      </td>
    </tr>
  <% end %>
  </tbody>
</table>

<br />

<hr />

<h1>Add new Task</h1>
<form method="POST" action="/tasks">
  <label for="name">Name:</label>
  <br />
  <input type="text" id="name" name="name">
  <br><br>

  <label for="description">Description:</label>
  <br>
  <textarea id="description" name="description">
  </textarea>
  <br><br>
```

```
  <input type="submit">
</form>
```

The preceding HTML template was taken from *Chapter 7*, where we list all the tasks in our system. We just updated the table body to use the `stringify_task/1` function that's responsible for printing each task. For this to work, we will need all the functions defined in the view module to be accessible at the time of evaluation of the template. We can do that by leveraging the `:functions` option that can be passed to the `EEx.eval_file/3` function. Let's first start off by adding a `__using__/1` macro to `Goldcrest.View`:

```
defmodule Goldcrest.View do
  defmacro __using__(opts) do
    quote do
      @opts unquote(opts)

      def __goldcrest_view_using_options__, do: @opts

      def render(conn, file, assigns) do
        Goldcrest.ResponseHelpers.render(
          conn,
          file,
          assigns
        )
      end
    end
  end
end
```

In the preceding code snippet, we defined a `__using__/1` macro in `Goldcrest.View`, which defines a `__goldcrest_view_using_options__/0` function similar to `__goldcrest_controller_using_options__/0` in `Goldcrest.Controller`. We also defined a `render/3` function that simply delegates to `Goldcrest.ResponseHelpers.render/3`.

To pass helper functions to the evaluation of templates, we will need to update the `render/3` clause that evaluates the template file in the `Goldcrest.ResponseHelpers` module:

```
defmodule Goldcrest.ResponseHelpers do
  # ..
  def render(conn, file, %{assigns: assigns, opts: opts}) do
    contents =
      file
      |> html_file_path()
      |> EEx.eval_file([assigns: assigns], opts)

    render(conn, :html, contents)
```

```
    end

  #..
end
```

In the preceding code snippet, we changed the generic function clause for `render/3`, where we can pass both the `assigns` keyword and additional options as `opts` to the `EEx.eval_file/3` call. This allows us to pass `:functions` as part of the `opts` Keyword list.

Next, we will be updating `Goldcrest.View` to call `render/3` with proper arguments when the second argument is a string. In this way, it will not interfere with other `render/3` clauses in the `Goldcrest.ResponseHelpers` module:

```
defmodule Goldcrest.View do
  defmacro __using__(opts) do
    quote do
      alias Goldcrest.ResponseHelpers

      @opts unquote(opts)

      def __goldcrest_view_using_options__, do: @opts

      def render(conn, file, assigns) when is_binary(file)
      do
        ResponseHelpers.render(
          conn,
          file,
          %{
            assigns: assigns,
            opts: [
              functions: helper_functions()
            ]
          }
        )
      end

      def render(conn, type, data) do
        ResponseHelpers.render(conn, type, data)
      end

      defp helper_functions do
        [{
          __MODULE__,
          :functions
          |> __MODULE__.__info__()
          |> List.delete(
```

```
              {:__goldcrest_view_using_options__, 0})
           |> List.delete({:render, 3})
         }]
       end
     end
   end
 end
```

In the preceding code snippet, we update `render/3` to call the newly updated `Goldcrest.ResponseHelpers.render/3` function when the second argument is a binary, which means it's a filename. We pass `assigns` and `opts` containing `:functions` as the third argument to the function. This allows us to evaluate the template by calling any functions returned by `helper_functions/0`.

We also added a private `helper_functions/0` function, which uses the `__MODULE__.__info__/1` reflection provided by Elixir to get a list of all non-private functions defined in the view module. We then filter out `__goldcrest_view_using_options__/0` and `render/3` as they'll not be used inside the template. In this way, any non-private function defined in a view module can be used in the corresponding template during evaluation.

Finally, we added a generic clause for `render/3` that simply delegates to `Goldcrest.ResponseHelpers.render/3` for other use cases.

This wraps up our `View` DSL. Now that we have both the `Controller` and `View` DSLs built, it's time to build a web app using these DSLs.

Building a web app using the DSLs

Let's take the web app we built in *Chapter 7* as a starting point. Before making any changes, let's ensure the app is still running properly by running the following command:

```
$ mix run --no-halt
```

Let's go to `http://localhost:4040/tasks` in our browsers to make sure the page is rendered as follows:

Figure 9.1: Expected page

At this point, the `TasksWeb.TaskController` module should look as follows:

lib/tasks_web/controller/task_controller.ex

```elixir
defmodule TasksWeb.TaskController do
  use Plug.Builder
  import Plug.Conn
  alias TasksWeb.Tasks

  def call(conn, action: action) do
    conn = super(conn, [])

    apply(__MODULE__, action, [conn, conn.params])
  end

  def index(conn, _params) do
    tasks = Tasks.list()

    conn
    |> put_status(200)
```

```
      |> render("tasks.html.eex", tasks: tasks)
    end

    # ..
  end
```

Let's remove the `render/3` and `call/2` functions with all the code besides the controller actions and replace it with `use Goldcrest.Controller`.

Our controller module should now look exactly as follows:

lib/tasks_web/controller/task_controller.ex

```
defmodule TasksWeb.TaskController do
  use Goldcrest.Controller

  alias TasksWeb.Tasks

  def index(conn, _params) do
    tasks = Tasks.list()

    conn
    |> put_status(200)
    |> render("tasks.html.eex", tasks: tasks)
  end

  def create(conn, %{"name" => name, "description" =>
    description}) do
    Tasks.add(name, description)

    conn
    |> redirect(to: "/tasks")
  end

  def delete(conn, %{"id" => id}) do
    id
    |> String.to_integer()
    |> Tasks.delete()

    conn
    |> redirect(to: "/tasks")
  end
end
```

Make sure any other references are removed.

Next, let's do the same for the `TasksWeb.TaskView` module, which starts as follows:

lib/tasks_web/view/task_view.ex

```
defmodule TasksWeb.TaskView do
  def render(conn, file, assigns) do
    Goldcrest.View.render(__MODULE__, conn, file, assigns)
  end
end
```

Let's update the view module such that it only has `use Goldcrest.View`:

lib/tasks_web/view/task_view.ex

```
defmodule TasksWeb.TaskView do
  use Goldcrest.View
end
```

While we're updating the view module, let's also add the `stringify_task/1` helper function to test whether functions are being properly passed to `EEx.eval_file/3` when evaluating templates:

lib/tasks_web/view/task_view.ex

```
defmodule TasksWeb.TaskView do
  use Goldcrest.View

  def stringify_task({name, description}) do
    "#{name} - #{description}"
  end
end
```

Now, the only thing that's missing is updating the `tasks.html.eex` template to utilize the `stringify_task/1` helper. Let's update the template to look like the following file:

priv/templates/tasks.html.eex

```
<h1>Listing Tasks</h1>
<table>
  <thead>
    <tr>
      <th>Task</th>
      <th></th>
    </tr>
  </thead>
  <tbody>
```

```
<%= for {task, index} <- Enum.with_index(@tasks) do %>
  <tr>
    <td><%= stringify_task(task) %></td>
    <td>
      <a href="/tasks/<%= index %>/delete">Delete</a>
    </td>
  </tr>
<% end %>
</tbody>
</table>

<br />

<hr />

<h1>Add new Task</h1>
<form method="POST" action="/tasks">
  <label for="name">Name:</label>
  <br />
  <input type="text" id="name" name="name">
  <br><br>

  <label for="description">Description:</label>
  <br>
  <textarea id="description" name="description">
  </textarea>
  <br><br>

  <input type="submit">
</form>
```

The preceding file updates the table to just have one column instead of two and lets stringify_
task/1 display the task in a user-friendly manner.

Now, let's restart our app again and test the entire functionality, from creating a task, listing tasks,
and deleting a task:

```
$ mix run --no-halt
```

Figure 9.2: Empty tasks

In the preceding screenshot, we start with zero tasks in our store. We then add a new task by adding a name and a description and clicking **Submit**:

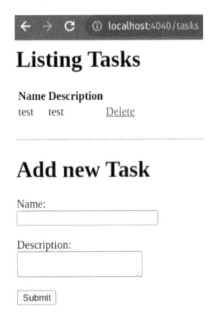

Figure 9.3: A new task added with a Delete link

We can see in the previous screenshot how we now have a new task in the table, with a delete link. Let's click on **Delete** next to the task:

Figure 9.4: After clicking the Delete link

After clicking on **Delete**, we can see that the task was successfully deleted.

After running the preceding tests, we can conclusively say that the web app is working with our new DSLs.

We have now learned how to simplify our Controller and View DSLs such that other developers can easily use them in their projects. However, as we learned in *Chapter 8*, any DSL is incomplete without allowing developers to easily test whether they're using the DSLs the way they should. In this case, we want to allow developers to easily test both the controllers and the views.

Creating test helpers for ease of testing and debugging

Let's start off by creating a `ControllerCase` module right away. This is where we will house all our helper functions, imports, and other behavior required for the tests:

```
defmodule Goldcrest.ExUnit.ControllerCase do
  defmacro __using__(_) do
    quote do
      use ExUnit.Case
      use Plug.Test
```

```
        end
      end
    end
```

In the preceding code snippet, we added a `__using__/1` macro, which uses two modules, `ExUnit.Case` and `Plug.Test`. We will need `ExUnit.Case` since we will be using `ExUnit` for tests. We will also need `Plug.Test` as seen in *Chapter 5*, as it provides several helpful functions, such as `conn/2`, `get_session/2`, and so on, which are needed to properly test a controller.

Let's start by writing a controller test for `TasksWeb.TaskController`:

```
defmodule TasksWeb.TaskControllerTest do
  use Goldcrest.ExUnit.ControllerCase

  describe "GET /tasks -- index/2" do
    test "renders a list of tasks" do
      conn = conn(:get, "/tasks")

      conn = TasksWeb.Router.call(conn, [])

      IO.inspect(conn.resp_body)

      assert true
    end
  end
end
```

In the preceding test module, we use the `Plug.Test.conn/2` function to initialize a `%Plug.Conn{}` struct, which we can further use while calling a plug module. For any controller test, the plug module will be the router, in this case, `TasksWeb.Router`. Let's run the preceding test and see what it prints as `conn.resp_body`:

```
$ mix test
.."<h1>Listing Tasks</h1>\
n<table>\n  <thead>\n    <tr>\n        <th>Task</th>\n
<th></th>\n    </tr>\n  </thead>\n  <tbody>\n  \n  </tbody>\n</
table>\n\n<br />
\n\n  <hr />\n\n<h1>Add new Task</h1>\n<form method=\"POST\" action=\"/
tasks\">\n
<label for=\"name\">Name:</label>\n  <br />\n  <input type=\"text\"
id=\"name\"
name=\"name\">\n  <br><br>\n\n  <label
for=\"description\">Description:</label>
\n <br>\n  <textarea id=\"description\" name=\"description\">\n  </
textarea>\n
<br><br>\n\n  <input type=\"submit\">\n</form>\n"
.
```

```
Finished in 0.04 seconds (0.00s async, 0.04s sync)
1 doctest, 2 tests, 0 failures
```

As we can see, the HTML body in `conn.resp_body` is what we'd expect it to be. We can see that the HTML body has no tasks because we haven't added any.

Now, let's add a task to `Task.Store` and rerun the test:

```
defmodule TasksWeb.TaskControllerTest do
  use Goldcrest.ExUnit.ControllerCase

  describe "GET /tasks -- index/2" do
    test "renders a list of tasks" do
      conn = conn(:get, "/tasks")

      # Adding a task
      name = "Test Name"
      description = "Test Description"
      Tasks.add(name, description)

      conn = TasksWeb.Router.call(conn, [])

      IO.inspect(conn.resp_body)

      assert true
    end
  end
end
```

In the preceding code snippet, we added a call to `Tasks.add/2` right before we made the call to the router plug. Let's run the test again:

```
$ mix test
.."<h1>Listing Tasks</h1>\
n<table>\n  <thead>\n    <tr>\n      <th>Task</th>\n
<th></th>\n    </tr>\n  </thead>\n  <tbody>\n  \n    <tr>\n
<td>Test Name - Test Description</td>\n      <td>\n
<a href=\"/tasks/0/delete\">Delete</a>\n        </td>\n    </
tr>\n  \n  </tbody>\n
</table>\n\n<br />\n\n<hr />\n\n<h1>Add new Task</h1>\n<form
method=\"POST\"
action=\"/tasks\">\n  <label for=\"name\">Name:</label>\n  <br
/>\n  <input
type=\"text\" id=\"name\" name=\"name\">\n  <br><br>\n\n  <label
for=\"description\">Description:</label>\n  <br>\n  <textarea
id=\"description\""
```

```
name=\"description\">\n  </textarea>\n  <br><br>\n\n  <input
type=\"submit\">
\n</form>\n"
```

.

```
Finished in 0.04 seconds (0.00s async, 0.04s sync)
1 doctest, 2 tests, 0 failures
```

We can see that now the table isn't empty, and it renders the newly added task. So, let's add an assertion to check that:

```
defmodule TasksWeb.TaskControllerTest do
  use Goldcrest.ExUnit.ControllerCase

  alias TasksWeb.Tasks

  describe "GET /tasks -- index/2" do
    test "renders a list of tasks" do
      conn = conn(:get, "/tasks")

      # Adding a task
      name = "Test Name"
      description = "Test Description"
      Tasks.add(name, description)

      conn = TasksWeb.Router.call(conn, [])

      assert conn.resp_body =~ "Listing Tasks"

      stringified_task = TasksWeb.TaskView.stringify_task({
        name, description})
      assert conn.resp_body =~ stringified_task
    end
  end
end
```

In the preceding code, we've added two assertions. The first one checks at a very high level whether the correct page is rendered by checking whether the heading exists. The second one, more specifically, checks whether the newly added tasks are visible in the rendered HTML.

We can further simplify the interface by wrapping up the calls to conn/2 and the router call in a single function. Let's define these helper functions in the Goldcrest.ExUnit.ControllerCase module:

```
defmodule Goldcrest.ExUnit.ControllerCase do
  defmacro __using__(_) do
    quote do
```

```
        use ExUnit.Case
        use Plug.Test
    end
  end

  def request(method, path, params_or_body \\ nil, router)

  def request(method, path, params, router) when
    is_map(params) do
    request(method, path, Jason.encode!(params), router)
  end

  def request(method, path, params_or_body, router) do
    conn = Plug.Test.conn(method, path, params_or_body)

    router.call(conn, [])
  end
end
```

We started by adding a `request/4` function that takes a method, a path, params, and a router, and returns a `%Plug.Conn{}` struct with the response attributes. We made the `request/4` function dynamic enough that we can pass both a map and an encoded body as the `params_or_body` argument. This allows us to write cleaner controller tests that rely on sending parameters as maps.

We're not done yet; to make this even more idiomatic, we should get rid of the parameters that can be inferred at compile time. In this case, we know that the router module will be the same for a controller test. Therefore, we can let a controller test pass the `:router` option while using the `ControllerCase` module. Similar to `Goldcrest.Controller`, we can also have a default router based on namespacing. Let's go ahead and add this to our module:

```
defmodule Goldcrest.ExUnit.ControllerCase do
  defmacro __using__(opts) do
    quote do
      use ExUnit.Case
      use Plug.Test

      @opts unquote(opts)

      @default_router_module __MODULE__
                             |> Module.split()
                             |> Enum.take(1)
                             |> List.insert_at(1, ".Router")
                             |> Module.concat()

      @router_module @opts[:router] ||
```

```elixir
      @default_router_module

    @helper_module unquote(__MODULE__)

    def get(path, params_or_body \\ nil) do
      @helper_module.request(:get, path, params_or_body,
                             @router_module)
    end

    def post(path, params_or_body \\ nil) do
      @helper_module.request(:post, path, params_or_body,
                             @router_module)
    end

    def put(path, params_or_body \\ nil) do
      @helper_module.request(:put, path, params_or_body,
                             @router_module)
    end

    def patch(path, params_or_body \\ nil) do
      @helper_module.request(:patch, path,
        params_or_body, @router_module)
    end

    def delete(path, params_or_body \\ nil) do
      @helper_module.request(:delete, path,
        params_or_body, @router_module)
    end
  end
end

def request(method, path, params_or_body \\ nil, router)

def request(method, path, params, router) when
  is_map(params) do
  request(method, path, Jason.encode!(params), router)
end

def request(method, path, params_or_body, router) do
  conn = Plug.Test.conn(method, path, params_or_body)

  router.call(conn, [])
end
end
```

In the preceding code snippet, we added the `get`, `post`, `put`, `patch`, and `delete` functions, each corresponding to an HTTP verb. All these functions simply delegate to the `Goldcrest.ExUnit.ControllerCase.request/4` function. We also allow developers to pass the `:router` parameter as an option while using `ControllerCase`, but default it to the topmost namespaced `Router` module, which is the convention. This allows us to simply call `get(path)` to test a GET request.

Now, let's use these newly created helper functions to test our controller thoroughly:

```
defmodule TasksWeb.TaskControllerTest do
  use Goldcrest.ExUnit.ControllerCase

  alias TasksWeb.Tasks

  describe "GET /tasks -- index/2" do
    setup do
      name = "Test Name"
      description = "Test Description"
      Tasks.add(name, description)

      {:ok, name: name, description: description}
    end

    test "renders a list of tasks", %{name: name,
      description: description} do
      conn = get("/tasks")

      assert conn.resp_body =~ "Listing Tasks"

      stringified_task = TasksWeb.TaskView.stringify_task({
        name, description})
      assert conn.resp_body =~ stringified_task
    end
  end
end
```

In the previous code snippet, we updated `TasksWeb.TaskControllerTest` to use the `get/1` function to test the index action. Let's test the `create/2` and `delete/2` actions next. Keep in mind it's always best to delete all data in the store between tests so they don't interfere with each other. So, we'll need to implement a function to clear the store as well between all tests:

```
defmodule TasksWeb.TaskControllerTest do
  use Goldcrest.ExUnit.ControllerCase

  alias TasksWeb.Tasks

  setup do
```

```
    clear_store()

    :ok
end

describe "GET /tasks -- index/2" do
  setup do
    name = "Test Name"
    description = "Test Description"
    Tasks.add(name, description)

    {:ok, name: name, description: description}
  end

  test "renders a list of tasks", %{name: name,
    description: description} do
    conn = get("/tasks")

    assert conn.resp_body =~ "Listing Tasks"

    stringified_task = TasksWeb.TaskView.stringify_task({
      name, description})
    assert conn.resp_body =~ stringified_task
  end
end

describe "POST /tasks -- create/2" do
  test "creates a new task and redirects to index" do
    name = "Test Name"
    description = "Test Description"

    stringified_task = TasksWeb.TaskView.stringify_task({
      name, description})

    # Task doesn't exist at first
    conn = get("/tasks")
    assert conn.resp_body =~ "Listing Tasks"
    refute conn.resp_body =~ stringified_task

    conn = post("/tasks", %{name: name,
      description: description})

    # Redirects to index
    assert conn.status == 302
```

```
      assert {"location", "/tasks"} in conn.resp_headers

      # Task exists on the index page
      conn = get("/tasks")
      assert conn.resp_body =~ stringified_task
    end
  end

  describe "GET /tasks/:id -- delete/2" do
    setup do
      name = "Test Name"
      description = "Test Description"
      Tasks.add(name, description)

      {:ok, name: name, description: description}
    end

    test "deletes the task", %{name: name, description:
      description} do
      conn = get("/tasks")

      assert conn.resp_body =~ "Listing Tasks"

      # Task appears in the list at first
      stringified_task = TasksWeb.TaskView.stringify_task({
        name, description})
      assert conn.resp_body =~ stringified_task

      conn = get("/tasks/1/delete")

      # Redirects to index
      assert conn.status == 302
      assert {"location", "/tasks"} in conn.resp_headers

      conn = get("/tasks")

      # Task doesn't appear in the list
      stringified_task = TasksWeb.TaskView.stringify_task({
        name, description})
      assert conn.resp_body =~ stringified_task
    end
  end

  defp clear_store do
```

```
      Agent.update(TasksWeb.Tasks, fn _ -> [] end)
    end
  end
```

We updated `TasksWeb.TaskControllerTest` to run `clear_store/0` before every test, ensuring a clean database. We also added test cases for the `create/2` action using the `post/2` call and the `delete/2` action using the `get/1` call. We can see that we're using a map as the request body for the `post/2` call, which is a lot easier to read than translating it into `json`.

We now have an idiomatic and concise way of testing our controllers. There are a lot more improvements that could be made to this, but those are beyond the scope of this chapter.

Summary

In this chapter, we first looked at our use case and evaluated whether it makes sense to use metaprogramming here. Since we're building a DSL with clear requirements and proper implementation already done, it was an easy choice to make the interface cleaner by adding a layer of metaprogramming.

Then, we designed our DSL based on what would provide the best experience for developers using our project. Once we had the requirements and design ready, we used the constructs of metaprogramming we learned in *Chapter 8* to build DSLs for both `Controller` and `View`.

Just as we learned in *Chapter 8*, building a DSL is only half a job done. We proceeded to add introspective features and test helpers to allow for easier adoption, testing, maintenance, and debugging. We also updated the web app from *Chapter 7* to work with the new DSL and ensured everything worked the way we expected it to.

Now that we have built a DSL for our `Controller` and `View` modules, the book's last chapter focuses on building a cleaner DSL for our `Router` module to create a synergy between all the components of Goldcrest.

Exercises

Test your knowledge by trying to implement the following:

- Build a test helper for the `View` module. Does it make sense to have test helpers for views? Which parts of the views need to be tested and why?

- Update `Goldcrest.ExUnit.ControllerCase.request/4` so that you can pass it a `%Plug.Conn{}` struct. This will help us to send `conn` with a specific setup, such as authorization headers or session data.

10
Building the Router DSL

In this chapter, we will continue using the metaprogramming skills learned in *Chapter 8* to make `Goldcrest` easier to use and adopt. In the previous chapter, we used metaprogramming to create a DSL around our controller and view modules. Doing this eliminated a lot of boilerplate code while significantly increasing the readability of those modules. We also added reflections and test helpers to allow us to test and debug those modules well. In this chapter, we will do the same for the router module. Similar to *Chapter 9*, we will start by defining the requirements for our router interface, see why metaprogramming makes sense for this use case, and finally, build the DSL along with high reflectivity and testability.

We will cover the following topics in this chapter:

- Designing the DSL for routers based on requirements

- Using Elixir's pillars of metaprogramming to build the router DSL

- Adding reflections to make it easier to debug and test the router DSL

- Updating the `tasks` app to use the router DSL

- Creating test helper modules for the router DSL

At the end of this chapter, we will have a functional DSL to define routers that work well with a controller and view the DSLs defined in the previous chapter, making it easier to define, test, and debug code related to a router.

Technical requirements

In this chapter, we go back to the example web application we built in *Part 2* of this book. We will look at the routers and router plugs in those applications, along with the controllers and views built in the previous chapter, and use those to define the requirements of the router DSL.

Similar to previous chapters, Elixir `1.12.x` and Erlang `23.2.x` are required to properly run the code snippets in this chapter. I'd also recommend coding along and understanding why we decided to use metaprogramming for this use case.

The code examples for this chapter can be found at `https://github.com/PacktPublishing/Build-Your-Own-Web-Framework-in-Elixir/tree/main/chapter_10`

Why use metaprogramming here?

Similar to *Chapter 9*, we will use metaprogramming here to wrap an already existing implementation of our router and router pipeline around a digestible and idiomatic interface.

Metaprogramming is especially useful in the case of the router modules because, currently, the router module is the least readable and most boilerplate-heavy of all components in Goldcrest. We also want our router to work well with all the controllers and views, so it puts an added burden on developers to make sure all the components are hooked up correctly. Therefore, by taking away the responsibility of managing that boilerplate code, we will significantly improve the developer experience of building a router using Goldcrest.

Now that we understand why the Goldcrest router is a good candidate for metaprogramming, let's start by defining the requirements for the router DSL.

DSL design

Similar to the previous chapter, we want to build our DSL in a way that maximizes its readability by only displaying information that's specific to what developers will be writing. The current version of the router in the `tasks` application looks like this:

```
defmodule TasksWeb.Router do
  use Plug.Router

  alias TasksWeb.TaskController

  plug Plug.Parsers,
    parsers: [:urlencoded, :multipart],
    pass: ["text/html", "application/*"]

  plug :parse_body
  plug :match
  plug :dispatch

  get "/tasks/:id/delete", do: TaskController.call(conn,
    action: :delete)
  get "/tasks", do: TaskController.call(conn, action:
    :index)
  post "/tasks", do: TaskController.call(conn, action:
    :create)
```

```
  match _ do
    send_resp(conn, 404, "<h1>Not Found</h1>")
  end

  defp parse_body(conn, _opts) do
    {:ok, body, conn} = Plug.Conn.read_body(conn,
      length: 1_000_000)

    body_params = decode_body_params(body)
    params = Map.merge(conn.params, body_params)
    conn = %{conn | body_params: body_params,
              params: params }

    conn
  end

  defp decode_body_params(""), do: %{}
  defp decode_body_params(binary),
    do: Jason.decode!(binary)
end
```

In the preceding router, we need the use `Plug.Router` line so that we can leverage the predefined plug, `Plug.Router`, which allows us to define a pipeline of plugs that handle a request by matching on a method and a path. We also need `Plug.Parsers`, which makes it easier for us to parse the body of a request by using built-in parsers for HTML and JSON. We also define a custom plug, `:parse_body`, which simply decodes the body of the conn struct, before calling the `match` and `dispatch` plugs, which are both needed for requests to be properly routed to a controller. Finally, we also have a default match clause to handle any requests that don't match explicitly defined routes. This allows us to properly handle 404 requests. Therefore, the only lines that are custom to a router are the routes that are defined using calls such as `get` and `post`. We also should give a router the ability to pass options to the `Plug.Parsers` plug for different parsing strategies.

After removing all the boilerplate code, our router DSL should look a lot simpler and digestible, as shown here:

```
defmodule TasksWeb.Router do
  use Goldcrest.Router,
    parser_options: [
      parsers: [:urlencoded, :multipart],
      pass: ["text/html", "application/*"]
    ]

  alias TasksWeb.TaskController

  get "/tasks/:id/delete", TaskController, :delete
```

```
    get "/tasks", TaskController, :index
    post "/tasks", TaskController, :create
  end
```

In the preceding code, we have limited the code that adds the router behavior to our module to the use statement. We also pass `parser_options`, which can simply be forwarded to the options for `Plug.Parsers`. We can also see that defining a route is a lot easier to read, as it simply involves a controller module and an action name, instead of calling the `TaskController.call/2` function.

Now that we have a goal set for our DSL, let's start building it.

Building the router DSL

In order for the use `Goldcrest.Router, ...` statement to work, we will first need to define a `__using__/1` macro in the `Goldcrest.Router` module. This macro will be responsible for injecting all the router behavior into a module.

Let's start by simply moving all the `plug` calls and private `parse_body` and `decode_body_params` functions to the `__using__/1` macro:

```
defmodule Goldcrest.Router do
  import Plug.Conn

  defmacro __using__(opts) do
    quote do
      use Plug.Router

      @opts unquote(opts)

      plug Plug.Parsers,
        parsers: [:urlencoded, :multipart],
        pass: ["text/html", "application/*"]

      plug :parse_body
      plug :match
      plug :dispatch

      defp parse_body(conn, _opts) do
        {:ok, body, conn} = Plug.Conn.read_body(conn,
          length: 1_000_000)

        body_params = decode_body_params(body)
        params = Map.merge(conn.params, body_params)
        conn = %{conn | body_params: body_params,
                  params: params }
```

```
          conn
       end

     defp decode_body_params(""), do: %{}
     defp decode_body_params(binary),
       do: Jason.decode!(binary)
   end
  end
end
```

In the preceding code snippet, we simply moved all the plug calls and private functions in our old router to the `using` macro, as they will mostly be the same for other routes too. One thing to keep in mind is that we will likely need to make the `Plug.Parsers` call more dynamic for requests that use different `Accept` headers.

Now, we can update the `TasksWeb.Router` module to work with the `Goldcrest.Router` `__using__/1` macro. Since we have extracted everything but the routes and match clauses, we can replace all of that code with the `use` statement:

```
defmodule TasksWeb.Router do
  use Goldcrest.Router

  alias TasksWeb.TaskController

  get "/tasks/:id/delete", do: TaskController.call(conn,
    action: :delete)
  get "/tasks", do: TaskController.call(conn,
    action: :index)
  post "/tasks", do: TaskController.call(conn,
    action: :create)

  match _ do
    send_resp(conn, 404, "<h1>Not Found</h1>")
  end
end
```

Now that we have access to options passed with the `use` statement call, we can use that to specify options to be used with the `Plug.Parsers` call and forward them to that plug. Let's update the `Plug.Parsers` call to use options from the `__using__/1` macro:

```
defmodule Goldcrest.Router do
  import Plug.Conn

  defmacro __using__(opts) do
    quote do
```

```
        use Plug.Router

        @opts unquote(opts)

        plug(Plug.Parsers, @opts[:parser_options])

        # ..
      end
    end
end
```

In the preceding code snippet, we updated the `Goldcrest.Router.__using__/1` macro to take `:parser_options` as a key in the `opts` Keyword list. We can now update `TasksWeb.Router` to send `parser_options`, along with the use call:

```
defmodule TasksWeb.Router do
  use Goldcrest.Router,
    parser_options: [
      parsers: [:urlencoded, :multipart],
      pass: ["text/html", "application/*"]
    ]

  alias TasksWeb.TaskController

  get "/tasks/:id/delete", do: TaskController.call(conn,
    action: :delete)
  get "/tasks", do: TaskController.call(conn,
    action: :index)
  post "/tasks", do: TaskController.call(conn,
    action: :create)

  match _ do
    send_resp(conn, 404, "<h1>Not Found</h1>")
  end
end
```

In the preceding code snippet, we still have the generic `match` statement, which is responsible for handling the requests that don't match any routes previously defined. However, we can't simply move the `match` statement in the `__using__/1` macro because it needs to be called after all the routes are defined.

Since we need to define the match clause at the very end of the router's compilation, we can use the @ before_compile callback. Let's register the @before_compile callback in the Goldcrest. Router.__using__/1 macro:

```
defmodule Goldcrest.Router do
  import Plug.Conn

  defmacro __using__(opts) do
    quote do
      use Plug.Router

      @opts unquote(opts)

      @before_compile
        unquote(__MODULE__).BeforeCompileHelpers

      # ..
    end
  end
end
```

Now that we have registered the @before_compile callback, let's define the __before_compile__/1 macro in the Goldcrest.Router.BeforeCompileHelpers module. This macro simply adds the generic match clause, which sends a 404 response:

```
defmodule Goldcrest.Router.BeforeCompileHelpers do
  defmacro __before_compile__(_macro_env) do
    quote do
      match _ do
        send_resp(conn, 404, "<h1>Not Found</h1>")
      end
    end
  end
end
```

Now, we can remove the generic match clause from TasksWeb.Router:

```
defmodule TasksWeb.Router do
  use Goldcrest.Router,
    parser_options: [
      parsers: [:urlencoded, :multipart],
      pass: ["text/html", "application/*"]
    ]

  alias TasksWeb.TaskController
```

```
   get "/tasks/:id/delete", do: TaskController.call(conn,
     action: :delete)
   get "/tasks", do: TaskController.call(conn,
     action: :index)
   post "/tasks", do: TaskController.call(conn,
     action: :create)
 end
```

In the preceding code snippet, we simplified the `TasksWeb.Router` module to work with `Goldcrest.Router`. However, if we try to compile the `tasks_web` application, we get a compilation error:

```
$ mix compile
==> goldcrest
Compiling 1 file (.ex)
==> tasks_web
Compiling 1 file (.ex)

== Compilation error in file lib/tasks_web/router.ex ==
** (CompileError) lib/tasks_web/router.ex:1: undefined function
match/2
    (goldcrest 0.1.0) expanding macro:
      Goldcrest.Router.BeforeCompileHelpers.__before_compile__/1
    lib/tasks_web/router.ex:1: TasksWeb.Router (module)
```

So, it seems we can't define a `match` clause in the `__before_compile__/1` macro. After taking a look at Plug's router module on GitHub, it was clear that macros such as `match`, `forward`, and `get` end up invoking a private function named `compile/5`, which is responsible for dynamically defining functions that correspond to a defined route. The `function` clause that finally ends up being defined is for a private function, `do_match/4`. This means if we simply define a function clause for `do_match/4` that takes a `Plug.Conn` struct as the first argument, as part of the `__before_compile__/1` macro, that should accomplish what a generic `match` statement does for the `TasksWeb.Router` module. So, let's update `Goldcrest.Router.BeforeCompileHelpers.__before_compile__/1` to define a generic clause for the `__before_compile__/1` macro:

```
defmodule Goldcrest.Router.BeforeCompileHelpers do
  defmacro __before_compile__(_macro_env) do
    quote do
      defp do_match(conn, _, _, _) do
        conn = send_resp(conn, 404, "<h1>Not Found</h1>")

        Plug.Router.__put_route__(conn, conn.request_path,
          fn conn, _ ->
          conn
        end)
      end
```

```
        end
      end
   end
```

Now, if we try to compile `tasks_web`, it should succeed:

```
$ mix compile
==> goldcrest
Compiling 1 file (.ex)
==> tasks_web
Compiling 1 file (.ex)
$
```

Now that we have removed the need to add a generic `match` statement to the router, the last thing left to build our DSL is simplifying the process of defining the routes. Currently, we use `get/3`, `post/3`, and so on, defined in the `Plug.Router` module to define each route, but for us to simplify it such that it takes a controller name and an action name, we will have to build a wrapper around those functions.

We can start by importing all the functions from `Plug.Router`, except `get/3`, `post/3`, and so on. We have to do this because `use  Plug.Router` also imports `Plug.Router`, so invoking another import with an `except` option will override the first import, called as part of `use  Plug.Router`:

```
defmodule Goldcrest.Router do
   import Plug.Conn

   defmacro __using__(opts) do
     quote location: :keep do
       use Plug.Router
       import Plug.Router, except: [get: 3, post: 3, put: 3,
         patch: 3, delete: 3]
       # ..
     end
   end
end
```

Now, we need to define new macros such as `get/3` and `post/3`, which will work with the new DSL. Let's define them in `Goldcrest.Router`, instead of defining them inside of the `__using__/1` macro. This way, we can simply import `Goldcrest.Router` inside the `Goldcrest.Router.__using__/1` macro, which will allow us to use the `get/3` and `post/3` macros in the `TasksWeb.Router` module:

```
defmodule Goldcrest.Router do
   import Plug.Conn

   defmacro __using__(opts) do
     quote location: :keep do
```

```
      use Plug.Router
      import Plug.Router, except: [get: 3, post: 3, put: 3,
        patch: 3, delete: 3]
      import unquote(__MODULE__)

      # ..
    end
  end

  defmacro get(path, controller, action) do
    quote location: :keep do
      Plug.Router.get(
        unquote(path),
        to: unquote(controller),
        init_opts: [
          action: unquote(action)
        ]
      )
    end
  end

  defmacro post(path, controller, action) do
    quote location: :keep do
      Plug.Router.post(
        unquote(path),
        to: unquote(controller),
        init_opts: [
          action: unquote(action)
        ]
      )
    end
  end
end
```

In the preceding code snippet, we defined the get/3 and post/3 macros, which take a path, a controller, and an action. Inside the quote block, they simply call Plug.Router.get/2 or Plug.Router.post/2 and delegate to the controller module as a plug. This invokes the call function in the controller module, with :init_opts set to [action: <action-name>]. We now have the get/3 and post/3 wrapper macros, which translate calls from a more idiomatic path-controller-action format to Plug.Router's preferred format. We can similarly define macros for put/3, patch/3, and delete/3.

Now that we have defined new macros to define routes, let's update our router to use these macros:

```
defmodule TasksWeb.Router do
  use Goldcrest.Router,
    parser_options: [
      parsers: [:urlencoded, :multipart],
      pass: ["text/html", "application/*"]
    ]

  alias TasksWeb.TaskController

  get "/tasks/:id/delete", TaskController, :delete
  get "/tasks", TaskController, :index
  post "/tasks", TaskController, :create
end
```

In the preceding code snippet, we managed to strip off most of the boilerplate code, and we can see the difference in readability and ease of maintenance when compared to the original router file at the beginning of this chapter.

Now that we have built our router DSL, it's time to add reflections to facilitate the ease of debugging, maintenance, and testing of routers defined using Goldcrest.

Adding reflections

In this section, we will add reflections to `Goldcrest.Router`. These functions will allow us to test, debug, and inspect our router DSL better.

Let's start by updating the `__using__/1` macro to define a reflection function that returns the options with which it was invoked, similar to the DSLs designed in the previous chapter:

```
defmodule Goldcrest.Router do
  import Plug.Conn

  defmacro __using__(opts) do
    quote do
      use Plug.Router

      @opts unquote(opts)

      # Add this line
      def __goldcrest_router_using_options__, do: @opts

      # ..
```

```
        end
      end
   end
```

In the preceding code, we defined a __goldcrest_router_using_options__/0 function that returns the compile-time value of the module attribute, @opts. As we can see in the preceding code, @opts gets assigned, using the unquote/1 macro, to the value of opts passed to the __using__/1 macro. This way, even after compilation, we can easily tell what options were passed to use the Goldcrest.Router call, by simply calling the __goldcrest_router_using_options__/0 function.

We can now tell at runtime what options are used with Goldcrest.Router. Now that we have added a simpler reflection, let's add a more complex function that allows us to inspect all the routes defined in a router at runtime.

Reflective routes

One thing that Phoenix's router allows us to do is introspect at runtime on what routes are defined in the router and what pipeline a request goes through for a route. To give the same level of introspection, we will also store all the routes defined using the Goldcrest.Router macros.

A great way of doing this is by using module attributes in a similar way to how our controller DSLs store all the defined plugs. So, let's update the get/3 and post/3 macros to define and update an accumulating module attribute, @__routes__:

```
defmodule Goldcrest.Router do
  import Plug.Conn

  defmacro __using__(opts) do
    quote location: :keep do
      use Plug.Router
      import Plug.Router, except: [get: 3, post: 3, put: 3,
        patch: 3, delete: 3]
      import unquote(__MODULE__)

      @before_compile
        unquote(__MODULE__).BeforeCompileHelpers

      @opts unquote(opts)

      def __goldcrest_router_using_options__, do: @opts

      Module.register_attribute(__MODULE__, :__routes__,
        accumulate: true)
```

```elixir
      plug(Plug.Parsers, @opts[:parser_options])
    end
  end

  defmacro get(path, controller, action) do
    quote location: :keep do
      Module.put_attribute(
        __MODULE__,
        :__routes__,
        {{:get, unquote(path)}, {unquote(controller),
          unquote(action)}}
      )

      Plug.Router.get(
        unquote(path),
        to: unquote(controller),
        init_opts: [
          action: unquote(action)
        ]
      )
    end
  end

  defmacro post(path, controller, action) do
    quote location: :keep do
      Module.put_attribute(
        __MODULE__,
        :__routes__,
        {{:post, unquote(path)}, {unquote(controller),
          unquote(action)}}
      )

      Plug.Router.post(
        unquote(path),
        to: unquote(controller),
        init_opts: [
          action: unquote(action)
        ]
      )
    end
  end
end
```

In the preceding code snippet, we updated the `__using__`/1 macro to define a module attribute that accumulates when `Module.put_attribute/3` is called. This allows us to call `Module.put_attribute/3` in the `get/3` and `post/3` macros. In those macros, we simply update the `@__routes__` module attribute in the `{{method, path}, {controller, action}}` format. This way, at the end of the compilation of a router module, we will have access to all the routes in the `@__routes__` module attribute.

Now, we can add a function to the `__before_compile__`/1 macro, which simply returns the state of `@__routes__` right at the end of the compilation:

```
defmodule Goldcrest.Router.BeforeCompileHelpers do
  defmacro __before_compile__(_macro_env) do
    quote do
      defp do_match(conn, _, _, _) do
        conn = send_resp(conn, 404, "<h1>Not Found</h1>")

        Plug.Router.__put_route__(conn, conn.request_path,
          fn conn, _ ->
          conn
        end)
      end

      def __goldcrest_router_routes__, do: @__routes__
    end
  end
end
```

We now have access to all the routes defined in a router by simply calling the `__goldcrest_router_routes__`/0 function. We can use this function to build a helper module for routes similar to Phoenix.

Building router helpers

One of the most used features of the Phoenix router is the helper functions defined in a router helper module. Phoenix generates this module when a router is compiled. However, to keep our compilation simple, we can decouple router helper modules from the router and define `Goldcrest.Router.Helpers`, which can be used to define a helper module:

```
defmodule Goldcrest.Router.Helpers do
  @moduledoc false

  defmacro __using__(opts) do
    quote do
      @opts unquote(opts)
```

```
        @default_router_module __MODULE__
                               |> Module.split()
                               |> Enum.take(1)
                               |> List.insert_at(1, ".Router")
                               |> Module.concat()

      @router_module @opts[:router] ||
        @default_router_module
    end
  end
end
```

In the preceding code snippet, we added the __using__/1 macro to the Goldcrest.Router.
Helpers module, which takes a :router option. The router defaults to a module based on the top
namespace of the current module, similar to Goldcrest.Controller with the view module.
Now, we will define a path/3 function, which takes a controller, an action, and a list of bindings and
returns the final path as a string. In order to simplify the compilation process, we can delegate path/3,
defined inside the __using__/1 macro, to a path/4 function, defined in the Goldcrest.
Router.Helpers module:

```
defmodule Goldcrest.Router.Helpers do
  @moduledoc false

  defmacro __using__(opts) do
    quote do
      @opts unquote(opts)

      @default_router_module __MODULE__
                             |> Module.split()
                             |> Enum.take(1)
                             |> List.insert_at(1, ".Router")
                             |> Module.concat()

      @router_module @opts[:router] ||
        @default_router_module

      def path(controller, action, bindings \\ []) do
        unquote(__MODULE__).path(
          @router_module,
          controller,
          action,
          bindings
        )
      end
    end
  end
```

```
    end

    def path(router, controller, action, bindings \\ []) do
      routes = router.__goldcrest_router_routes__()

      route =
        Enum.find(routes, fn
          {{_action, _path}, {^controller, ^action}} -> true
          _ -> false
        end)

      apply_bindings(route, bindings)
    end

    defp apply_bindings({{_action, path}, {_, _}}, bindings)
    do
      Enum.reduce(bindings, path, fn {key, value}, acc ->
        String.replace(acc, ":#{key}", to_string(value))
      end)
    end
  end
end
```

In the preceding code snippet, we added a `path/3` function that gets defined for a router module and calls the `path/4` function, with the router module as the first argument. In the `Goldcrest.Router.Helpers.path/4` function, we use the `__goldcrest_router_routes__/0` reflection function, to get all the routes defined for the given router. We then use the list of routes to find a matching route for the given controller and action. Finally, we replace any path variables with the given bindings.

Next up, we can define the router helper module when the router module is compiled:

```
defmodule Goldcrest.Router do
  import Plug.Conn

  defmacro __using__(opts) do
    caller = __CALLER__.module

    quote location: :keep do
      use Plug.Router
      import Plug.Router, except: [get: 3, post: 3, put: 3,
        patch: 3, delete: 3]
      import unquote(__MODULE__)

      @before_compile
        unquote(__MODULE__).BeforeCompileHelpers
```

```
      @opts unquote(opts)

      def __goldcrest_router_using_options__, do: @opts

      Module.register_attribute(__MODULE__, :__routes__,
        accumulate: true)

      plug(Plug.Parsers, @opts[:parser_options])

      plug(:parse_body)
      plug(:match)
      plug(:dispatch)

      def parse_body(conn, _opts) do
        {:ok, body, conn} = read_body(conn, length:
          1_000_000)

        body_params = decode_body_params(body)
        params = Map.merge(conn.params, body_params)
        conn = %{conn | body_params: body_params,
          params: params}

        conn
      end

      defp decode_body_params(""), do: %{}
      defp decode_body_params(binary),
        do: Jason.decode!(binary)

      defmodule Helpers do
        use Goldcrest.Router.Helpers,
          router: unquote(caller)
      end
    end
  end
end

defmacro get(path, controller, action) do
  quote location: :keep do
    Module.put_attribute(
      __MODULE__,
      :__routes__,
      {{:get, unquote(path)}, {unquote(controller),
        unquote(action)}}
```

```
        )

        Plug.Router.get(
          unquote(path),
          to: unquote(controller),
          init_opts: [
            action: unquote(action)
          ]
        )
      end
    end

    defmacro post(path, controller, action) do
      quote location: :keep do
        Module.put_attribute(
          __MODULE__,
          :__routes__,
          {{:post, unquote(path)}, {unquote(controller),
            unquote(action)}}
        )

        Plug.Router.post(
          unquote(path),
          to: unquote(controller),
          init_opts: [
            action: unquote(action)
          ]
        )
      end
    end
  end
end
```

In the preceding code snippet, we defined a new module named Helpers, which will be under the namespace of the final router module. This module calls use Goldcrest.Router.Helpers with the given router. We also use the __CALLER__/0 macro outside the quote to get the name of the router module, which we finally use in the use statement.

Now, let's re-compile the tasks_web application and run an iex shell to test out the router helper module:

```
$ iex -S mix
iex> TasksWeb.Router.Helpers.path(TasksWeb.TaskController, :index)
"/tasks"
iex> TasksWeb.Router.Helpers.path(TasksWeb.TaskController, :delete)
"/tasks/:id/delete"
iex> TasksWeb.Router.Helpers.path(TasksWeb.TaskController, :delete,
```

```
id: 1)
"/tasks/1/delete"
```

We can see that all the routes are properly working. We even saw that for routes such as :delete that rely on bindings, the helpers default to path variables but apply those bindings when a value is given.

Now, we can update the Goldcrest.View module to import the router helper module, so we can use the path helper functions in our templates:

```
defmodule Goldcrest.View do
  defmacro __using__(opts) do
    quote do
      alias Goldcrest.ResponseHelpers

      @opts unquote(opts)

      @default_router_module __MODULE__
                             |> Module.split()
                             |> List.update_at(-1, fn str ->
                               String.replace(str,
                                 "Controller", "Router")
                             end)
                             |> Module.concat()

      @router_module @opts[:router_module] ||
        @default_router_module

      import @router_module.Helpers, as: Routes
    end
  end
end
```

```
Finally, we can update the index template to use Routes.path/3 helper
function instead of hard-coded routes.
<h1>Listing Tasks</h1>
<table>
  <thead>
    <tr>
      <th>Task</th>
      <th></th>
    </tr>
  </thead>
  <tbody>
  <%= for {task, index} <- Enum.with_index(@tasks) do %>
    <tr>
      <td><%= stringify_task(task) %></td>
      <td>
```

```
            <a href="<%= Routes.path(TaskWeb.TaskController,
               :delete, id: index) %>">Delete</a>
        </td>
      </tr>
   <% end %>
   </tbody>
</table>

<br />

<hr />

<h1>Add new Task</h1>
<form method="POST" action="<%= Routes.path(TaskWeb.TaskController,
:create) %>">
   <label for="name">Name:</label>
   <br />
   <input type="text" id="name" name="name">
   <br><br>

   <label for="description">Description:</label>
   <br>
   <textarea id="description" name="description">
   </textarea>
   <br><br>

   <input type="submit">
</form>
```

Now, let's restart our app again and test out the entire functionality, including creating a task, listing tasks, and deleting a task:

```
$ mix run --no-halt
```

Figure 10.1: Empty tasks

In the previous screenshot, we start with zero tasks in our store. We then add a new task by adding a name and a description and clicking **Submit**.

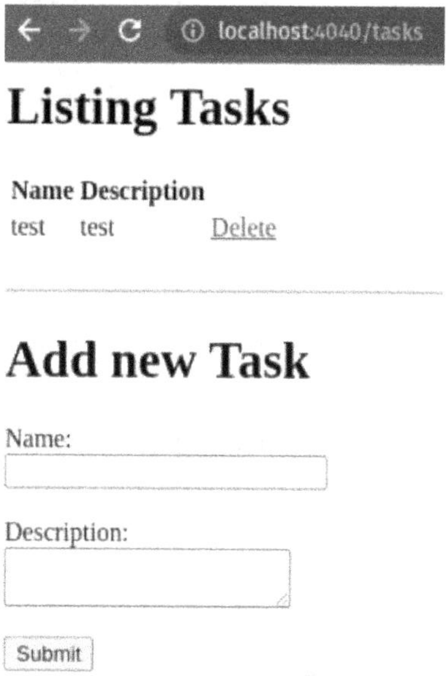

Figure 10.2: A new task added with the Delete link

We can see in the previous screenshot that we now have a new task in the table, with a **Delete** link. Let's click on **Delete**, next to the task.

Figure 10.3: After clicking the Delete link

After clicking on **Delete** in the previous screenshot, we can see that the task was successfully deleted.

After visiting the preceding URLs and performing those actions, we can conclusively say that the web app is working with the new router DSL and router helper module.

Now that we have a fully functional router DSL, it's time to add test helpers that will allow us to better test our routes.

Creating test helpers

One thing I always felt was missing from Phoenix's testing framework was the ability to simply test whether a set of routes is defined in a router. With our reflection function, it should be relatively easy to add that ability to Goldcrest.

Let's start by defining a `RouterCase` module that will house the logic for the `assert_route_defined?` function. We want to be able to call this function in the most idiomatic way possible; therefore, we will write this function to work with a given router, set as a module attribute in the `__using__/1` macro. This means a test module using `RouterCase` will be a singleton to one router. Let's go ahead and define that macro:

```
defmodule Goldcrest.ExUnit.RouterCase do
  defmacro __using__(opts) do
    quote do
```

```
       use ExUnit.Case

       @opts unquote(opts)

       @default_router_module __MODULE__
                              |> Module.split()
                              |> Enum.take(1)
                              |> List.insert_at(1, ".Router")
                              |> Module.concat()

       @router_module @opts[:router] ||
         @default_router_module

       @helper_module unquote(__MODULE__)

       def assert_route_defined?(method, match, controller,
         action) do
         routes =
           @router_module.__goldcrest_router_routes__()
         expected_route = {{method, match}, {controller,
           action}}

         ExUnit.Assertions.assert(
           expected_route in routes,
           """
           Expected the following route to be defined in the
           router, but it was not:

           #{@helper_module.humanize_route(expected_route)}.

           Defined routes:
           #{@helper_module.humanize_routes(routes)}
           """
         )
       end
     end
   end
end

def humanize_routes(routes) when is_list(routes) do
  routes
  |> Enum.map(&humanize_route/1)
  |> Enum.join("\n")
end

def humanize_route({{method, match}, {controller,
```

```
          action}}) do
      "#{inspect(method)} #{match}, #{controller}, #{action}"
    end
  end
```

In the preceding code snippet, we defined a `Goldcrest.ExUnit.RouterCase` module, which defines a `__using__`/1 macro. This macro sets a router module for the entire test module. This macro also adds ExUnit capabilities to the router test module and also defines the `assert_route_defined?`/4 function, which checks whether a route is defined for the set router with the given HTTP method, path, controller, and action. We use `ExUnit.Assertions.assert`/2 to add ExUnit capabilities to the `assert_route_defined?`/4 function, and we also add a custom failure message that lists all the routes defined in the given router.

Now that we have a `RouterCase` module defined, let's use it to test the `TasksWeb.Router` module:

```
defmodule TasksWeb.RouterTest do
  use Goldcrest.ExUnit.RouterCase

  alias TasksWeb.TaskController

  describe "routes" do
    test "defines GET /tasks routed to
      TaskController#index" do
      assert_route_defined?(:get, "/tasks", TaskController,
        :index)
    end

    test "defines GET /tasks/:id/delete routed to
      TaskController#delete" do
      assert_route_defined?(:get, "/tasks/:id/delete",
        TaskController, :delete)
    end

    test "defines POST /tasks routed to
      TaskController#create" do
      assert_route_defined?(:post, "/tasks",
        TaskController, :create)
    end
  end
end
```

In the preceding code snippet, we tested whether all the routes are defined in the given router. Each test case is simple and easy to read, ensuring that if a route is changed, we have a breaking test suite.

In order to run our tests, we can simply run the following command from the `tasks_web` folder:

```
$ mix test
......
Finished in 0.07 seconds (0.00s async, 0.07s sync)
6 tests, 0 failures
```

If we add a route that doesn't exist in the system, it will yield an appropriate test failure. Let's consider the following code snippet:

```
defmodule TasksWeb.RouterTest do
  use Goldcrest.ExUnit.RouterCase

  alias TasksWeb.TaskController

  describe "routes" do
    test "defines GET /tasks routed to
      TaskController#index" do
      assert_route_defined?(:get, "/tasks", TaskController,
        :index)
    end

    test "defines GET /tasks/:id/delete routed to
      TaskController#delete" do
      assert_route_defined?(:get, "/tasks/:id/delete",
        TaskController, :delete)
    end

    test "defines POST /tasks routed to
      TaskController#create" do
      assert_route_defined?(:post, "/tasks",
        TaskController, :create)
    end

    test "(expected to fail) POST /bad-route" do
      assert_route_defined?(:post, "/bad-route",
        TaskController, :bad_route)
    end
  end
end
```

Now, if we run the test suite, we will get a test failure with a custom message, as configured in the `RouterCase` module:

```
$ mix test
....


  1) test routes (expected to fail) POST /bad-route (TasksWeb.
RouterTest)
     test/tasks_web/router_test.exs:19
     Expected the following route to be defined in the router, but it
was
     not:

     :post /bad-route, Elixir.TasksWeb.TaskController, bad_route.

     Defined routes:
     :post /tasks, Elixir.TasksWeb.TaskController, create
     :get /tasks, Elixir.TasksWeb.TaskController, index
     :get /tasks/:id/delete,
        Elixir.TasksWeb.TaskController, delete

     code: assert_route_defined?(:post, "/bad-route",
        TaskController, :bad_route)
     stacktrace:
        test/tasks_web/router_test.exs:20: (test)


  ..
Finished in 0.07 seconds (0.00s async, 0.07s sync)
7 tests, 1 failure

Randomized with seed 299996
```

We now have an idiomatic, replicable, and concise way of defining, testing, and inspecting routers using the `Goldcrest` framework.

Summary

In this chapter, we first defined the desired DSL for our router module. We also understood why we use metaprogramming here, similar to the previous chapter.

Then, we took a look at the state of our router before this chapter and identified the lines that could be moved out of the router to the DSL. We started to build our DSL by moving all the `plug` calls to the `__using__/1` macro in `Goldcrest.Router`. We then found a way to add a generic `match` clause to get a default `404` response. We made the `use` statement dynamic by allowing it to pass options to the `Plug.Parsers` call as part of the `use` call.

We then worked on adding reflections and introspective features to the router DSL, similar to what we did in *Chapter 9* for controllers and views. We also defined helper functions and helper modules that allow us to reference a path by using a controller and action, similar to Phoenix. Finally, we added test helpers along with an example test for a router.

This concludes our minimal web framework, Goldcrest. We learned how to build our own HTTP server, how to use plugs to use the server effectively, and how controllers, views, templates, and routers work in Phoenix, building similar tools from the ground up. Finally, we learned how to wrap those into DSLs using metaprogramming. There are still many features that we can add to Goldcrest that are analogous to Phoenix, such as `LiveView`, better HTML support using `Phoenix.HTML`, and smarter EEx parsing using HEEx. We can also add more test helper modules to make testing even easier. Overall, my goal in writing this book was to somehow demystify modern-day web frameworks for software engineers and learn more about Elixir while doing so. A great next step after this book is to check out Phoenix's code base on GitHub and build more functionality on top of Goldcrest on your own. Hopefully, this hands-on book was able to reduce the barrier of entry for some developers so that they can start contributing to libraries such as Phoenix.

Exercises

Test your knowledge by implementing the following:

- What other test helpers could we add to `Goldcrest.ExUnit.RouterCase`? How would those tests help us build a more deterministic web application?

- Currently, the generic `match` clause only handles a `404` with HTML response. How can we update it to handle other types of responses?

Index

W

Packtpub.com

Subscribe to our online digital library for full access to over 7,000 books and videos, as well as industry leading tools to help you plan your personal development and advance your career. For more information, please visit our website.

Why subscribe?

- Spend less time learning and more time coding with practical eBooks and Videos from over 4,000 industry professionals

- Improve your learning with Skill Plans built especially for you

- Get a free eBook or video every month

- Fully searchable for easy access to vital information

- Copy and paste, print, and bookmark content

Did you know that Packt offers eBook versions of every book published, with PDF and ePub files available? You can upgrade to the eBook version at packtpub.com and as a print book customer, you are entitled to a discount on the eBook copy. Get in touch with us at customercare@packtpub.com for more details.

At www.packtpub.com, you can also read a collection of free technical articles, sign up for a range of free newsletters, and receive exclusive discounts and offers on Packt books and eBooks.

Packt is searching for authors like you

If you're interested in becoming an author for Packt, please visit `authors.packtpub.com` and apply today. We have worked with thousands of developers and tech professionals, just like you, to help them share their insight with the global tech community. You can make a general application, apply for a specific hot topic that we are recruiting an author for, or submit your own idea.

Hi!

I, Adi Iyengar, author of *Build Your Own Web Framework in Elixir*, really hope you enjoyed reading this book and found it useful for increasing your productivity and efficiency in Elixir and Phoenix.

It would really help us (and other potential readers!) if you could leave a review on Amazon sharing your thoughts on the book and what you learned from it.

Go to the link below or scan the QR code to leave your review:

`https://packt.link/r/1801812543`

Your review will help us to understand what's worked well in this book, and what could be improved upon for future editions, so it really is appreciated.

Best wishes,

Adi Iyengar

Download a free PDF copy of this book

Thanks for purchasing this book!

Do you like to read on the go but are unable to carry your print books everywhere?

Is your eBook purchase not compatible with the device of your choice?

Don't worry, now with every Packt book you get a DRM-free PDF version of that book at no cost.

Read anywhere, any place, on any device. Search, copy, and paste code from your favorite technical books directly into your application.

The perks don't stop there, you can get exclusive access to discounts, newsletters, and great free content in your inbox daily

Follow these simple steps to get the benefits:

1. Scan the QR code or visit the link below

https://packt.link/free-ebook/9781801812542

2. Submit your proof of purchase

3. That's it! We'll send your free PDF and other benefits to your email directly

www.ingramcontent.com/pod-product-compliance
Lightning Source LLC
Chambersburg PA
CBHW060529060326
40690CB00017B/3436